Transfrontier Conservation Areas

The introduction of transfrontier conservation areas (TFCAs) in Southern Africa was based on an enchanting promise: simultaneously contributing to global biodiversity conservation initiatives, regional peace and integration, and the sustainable socio-economic development of rural communities. Cross-border collaboration and eco-tourism became seen as the vehicles of this promise, which would enhance regional peace and stability along the way. However, as these highly political projects take shape, conservation and development policy making progressively shifts from the national to regional and global arenas, and the peoples most affected by TFCA formation tend to disappear from view.

This book focuses on the forgotten people displaced by, or living on the edge of, protected wildlife areas. It moves beyond the grand 'enchanting promise' of conservation and development across frontiers, and unfounded notions of TFCAs as integrated social-ecological systems. Peoples' dependency on natural resources – the specific combination of crop cultivation, livestock keeping and natural resource harvesting activities – varies enormously along the conservation frontier, as does their reliance on resources on the other side of the conservation boundary. Hence, the studies in this book move from the dream of eco-tourism-fuelled development supporting nature conservation and people towards the local realities facing marginalized people, living adjacent to protected areas in environments often poorly suited to agriculture.

Jens A. Andersson is a rural development sociologist who has worked on smallholder farming and migration in Southern Africa. He coordinates the 'Competing Claims on Natural Resources' programme, a collaboration between Wageningen University and several universities in Southern Africa.

Michel de Garine-Wichatitsky is an ecologist and a veterinarian who has worked on livestock–wildlife interactions in Southern Africa. He coordinates the collaborative research platform 'Production and Conservation in Partnership'.

David H.M. Cumming is an ecologist who has been working in conservation in Southern Africa since the early 1960s. He is presently an Honorary Professor at the University of Cape Town, a Research Associate at the University of Zimbabwe, a freelance consultant, and advisor to the AHEAD-GLTFCA initiative.

Vupenyu Dzingirai is a social anthropologist based at the Centre for Applied Social Sciences at the University of Zimbabwe. He has worked intensively in the Zambezi Valley among indigenous communities threatened by development activities.

Ken E. Giller is a Professor at Wageningen University (Netherlands), working principally on sustainable intensification of smallholder farming systems in sub-Saharan Africa. He is leader of the 'Competing Claims on Natural Resources' programme.

Transfrontier Conservation Areas

People living on the edge

Edited by
Jens A. Andersson,
Michel de Garine-Wichatitsky,
David H.M. Cumming,
Vupenyu Dzingirai
and Ken E. Giller

Routledge
Taylor & Francis Group
LONDON AND NEW YORK

earthscan
from Routledge

This first edition published 2013
by Routledge
2 Park Square, Milton Park, Abingdon, Oxon, OX14 4RN

Simultaneously published in the USA and Canada
by Routledge
711 Third Avenue, New York, NY 10017

Routledge is an imprint of the Taylor & Francis Group, an informa business

British Library Cataloguing in Publication Data
A catalogue record for this book is available from the British Library

Library of Congress Cataloging-in-Publication Data
Tranfrontier conservation areas: people living on the edge/edited by Jens A. Andersson . . . [et al.].
p. cm.
Includes bibliographical references and index.
1. Protected areas – Africa, Southern. 2. Transfrontier conservation areas – Africa, Southern. 3. Wildlife conservation – Africa, Southern.
I. Andersson, Jens A.
S944.5.P78T74 2013
333.720968 – dc23
2012011077

ISBN: 978-1-84971-208-8 (hbk)
ISBN: 978-1-84977-724-7 (ebk)

Typeset in Times by
Florence Production Ltd, Stoodleigh, Devon

Contents

Illustrations

Figures

Tables

Boxes

Contributors

Jens A. Andersson is a rural development sociologist working on rural–urban migration, rural livelihoods and the social organization of smallholder farming and informal trade in Eastern and Southern Africa. Andersson was regional coordinator of the interdisciplinary research programme 'Competing Claims on Natural Resources: Overcoming Mismatches in Resource Use from a Multi-Scale Perspective', a research fellow in the Plant Production Systems Group of Wageningen University, The Netherlands, and research associate of the Centre for Applied Social Sciences, University of Zimbabwe.

Frédéric Baudron is a tropical agronomist working for the International Maize and Wheat Improvement Centre (CIMMYT) in Addis Ababa, Ethiopia. He has worked for both agricultural and biodiversity conservation projects, and at various levels of the research development continuum. His Ph.D., conducted with Wageningen University and the research platform 'Production and Conservation in Partnership', explored the potential for agricultural intensification to save space for wildlife in northern Zimbabwe. His research interests include farming system analysis, sustainable intensification, and participatory innovation development.

Alexandre Caron's current focus in the research unit 'Animal et Gestion Intégrée des Risques' of CIRAD is the ecology of infectious disease transmission at the wildlife/domestic interface. Shifting the standard veterinarian approach from a disease-centred or a host-centred focus to a process-oriented focus, this area of research explores the properties of transmission processes using behavioural ecology, community ecology and molecular epidemiology. He is currently based in Harare, attached to the research platform 'Production and Conservation in Partnership'.

Petronella Chaminuka is an agricultural and natural resource economist at the University of Limpopo in South Africa. She has experience in research and development, and has worked with a variety of research institutions and non-governmental organizations. Her Ph.D. research analysed options for land use development in rural communities residing at the periphery of protected areas in South Africa. She also explored the socio-economic challenges associated with reconciling rural development and conservation goals. Her research interests include poverty analysis, land use modelling and participatory research for development.

Chaka Chirozva's current research interests are on innovative and decentralized knowledge generation approaches that promote inclusion of marginalized groups in governance of natural resources in areas such as game reserves (national parks), conservancies and transfrontier conservation areas. He is currently a lecturer at Bindura University where he teaches undergraduate courses in Social Anthropology and Organizational Behaviour.

David H.M. Cumming was previously Chief Ecologist and Deputy Director of Zimbabwe's Department of National Parks and Wild Life Management and later a senior conservation advisor for WWF. He is presently an honorary professor at the Percy FitzPatrick Institute, University of Cape Town, a research associate at the University of Zimbabwe, and an independent consultant in conservation and development. Current research interests include large herbivore impacts on biodiversity and ecosystem processes, and resilience and sustainability of transfrontier conservation areas.

Vupenyu Dzingirai graduated with a Ph.D. in 1998 at the University of Zimbabwe. He is an anthropologist with long-term experience in the Zambezi Valley with its conflicting development and conservation projects. His primary research interests include natural resource conflict and human mobility at various scales. Presently, he is based at the Centre for Applied Social Sciences, a communal resource management think tank at the University of Zimbabwe.

Hervé Fritz is Director of Research for CNRS and is based at University of Lyon 1, France. His main research interests are behavioural ecology, community ecology and ecosystem ecology. He has run research programmes in Southern Africa for 19 years and particularly in Zimbabwe. He is also Director of the Hwange Long Term Ecological Research site (LTER) studying the functioning of the broad socio-ecological system, including Hwange National Park and the Vice Chairman of the French–Zimbabwean research platform 'Production and Conservation in Partnership'.

Michel de Garine-Wichatitsky is a tropical ecologist and a veterinarian, working as Senior Researcher for CIRAD, based in Harare. His current research interests include ecological and epidemiological interactions between wild and domestic herbivores, and management of wild–domestic interfaces in the context of transfrontier conservation areas in Southern Africa. He is the coordinator of the research platform 'Production and Conservation in Partnership' and CIRAD representative in Zimbabwe.

Ken E. Giller is Professor of Plant Production Systems at Wageningen University. He leads a group of scientists with profound experience in systems analysis and simulation modelling scenarios of change. Ken's research has focused on smallholder farming systems in sub-Saharan Africa, and in particular problems of soil fertility and the role of nitrogen fixation in tropical legumes. He was scientific leader of the interdisciplinary project 'Competing Claims on Natural Resources: Overcoming Mismatches in Resource Use from a Multi-Scale Perspective' (www.competingclaims.nl/).

Chloé Guerbois is a junior researcher in ecology and socio-ecology at the National Natural History Museum in Paris, France. She is finishing her Ph.D. on integrated and sustainable wildlife management through protected areas in collaboration with the CNRS within the Hwange Long Term Ecological Research site (LTER) and the research platform 'Production and Conservation in Partnership'. Her work is based on the development of multidisciplinary approaches aimed at understanding and integrating socio-ecological constraints affecting resources uses and dynamics at the edge of Hwange National Park.

Eléonore Hellard is a junior ecologist with a multidisciplinary approach. After studying primates' behaviour and conservation in South Africa and Kenya, she investigated the ecological and anthropogenic determinants of human-spotted hyaena conflicts in periphery of Hwange National Park in Zimbabwe, in connection with the research platform 'Production and Conservation in Partnership'. She is currently based in the University of Lyon 1, France, where she completed her Ph.D. in eco-epidemiology and focuses on the detection and consequences of multiple infections in natural populations.

Ferran Jori is a veterinarian graduated at the Autonomous University of Barcelona (UAB) with 22 years of international experience in Africa and Latin America in the fields of rural development, wildlife management and epidemiology. Employed by CIRAD since 1998, the last eight years on the study of infectious diseases at the wildlife–livestock interface. Since 2007, he is based at the University of Pretoria participating in several projects on this topic in the Southern African region.

Collen Matema is a Ph.D. candidate at the Centre for Applied Social Sciences at the University of Zimbabwe, supported by the research platform 'Production and Conservation in Partnership'. He is investigating wildlife utilization potential for poverty reduction in the face of climate change in marginal areas of Zimbabwe. His research interests are centred on poverty, livelihoods and adaptation especially in economically marginal environments adjacent to wildlife areas.

Steven Matema is a social ecologist and Ph.D. student with Wageningen University, the Netherlands. His Ph.D. is on competing claims on natural resources, studying the socio-ecological and political dynamics of natural resource conflicts in Mbire District, one of the oldest CAMPFIRE areas in Zimbabwe. His research interests are on the institutional and social dynamics of development interventions in Zimbabwe, rural livelihood strategies, conservation and governance in and around wildlife areas.

Jessica Milgroom's research interests include population resettlement, policy in practice, access to natural resources, farming systems, and the role of action research in development. She carried out extensive fieldwork on the resettlement of villages from the Limpopo National Park in southern Mozambique for her recently completed Ph.D. Previously she worked in Brazil, Mexico, Côte d'Ivoire and Spain on diverse topics in agricultural development including agroforestry and soil erosion.

Billy Mukamuri is a lecturer at the Centre for Applied Social Sciences (CASS), University of Zimbabwe teaching postgraduate courses institutions and natural resource management. He is currently the chair of CASS, and leads the thematic field on natural resources governance and institutions within the research platform 'Production and Conservation in Partnership'. He is conducting field research on social change and community development in marginal areas close to protected areas.

Chrispen Murungweni's research interests are on the contribution of livestock and crop to sustainable poverty alleviation programmes that aim to improve livelihoods of the rural poor. He is interested in attaining these research objectives through the process of agriculture research for development. He recently completed a Ph.D. in the 'Competing Claims on Natural Resources' programme in Wageningen University, and is now employed by the public service commission of Zimbabwe, under the Ministry of Agriculture and Mechanisation in the Department of Research and Specialist Services (DR&SS), stationed at Grasslands Research Institute in Marondera.

Amon Murwira is a spatial ecologist and geographic information scientist based in the Department of Geography and Environmental Science at the University of Zimbabwe in Harare. He leads the thematic field on conservation and agriculture within the research platform 'Production and Conservation in Partnership'. His research interests include the use of earth observation technology, particularly remote sensing in combination with other sciences to understand spatial processes that shape landscapes where human and wildlife are the main land uses.

Tendai Nzuma is a postgraduate student at the National University of Science and Technology, supported by the research platform 'Production and Conservation in Partnership'. He has interests in studying the behaviour and life processes of animals and wildlife. For the past few years, he has been specializing in wildlife research and management, including the collection and analysis of biological data to determine the ecological effects of present and potential use of land through CBNRM programmes like CAMPFIRE in Zimbabwe.

Davies M. Pfukenyi is a senior lecturer and veterinary epidemiologist based in the Department of Clinical Veterinary Studies, University of Zimbabwe in Harare. He collaborates with the thematic field on animal health and environment within the research platform 'Production and Conservation in Partnership'.

His research interests are focused on the epidemiology of zoonoses at the domestic–wildlife interface and vector-borne parasitic diseases. He is also interested in research on quality assurance control and systems for food safety.

Xavier Poshiwa is a researcher based at Grasslands Research Institute, Marondera, Zimbabwe and currently a Ph.D. candidate in the 'Competing Claims on Natural Resources' programme in Wageningen University. Xavier's research for the past eleven years focused on livestock nutrition, fodder and pastures development and rangeland management for the smallholder and small-scale commercial farming systems of Zimbabwe. His main research interests are nature conservation, pastures and rangeland management and livestock nutrition.

Wayne Twine is a senior lecturer in the School of Animal, Plant and Environmental Sciences at the University of the Witwatersrand, South Africa. His research focuses on interactions between people and the environment in common-property resource systems. This includes the investigation of natural resource use, rural livelihoods, resource ecology and local resource management in the former apartheid homelands of South Africa.

Fadzai Zengeya's research interest is in spatial ecology, with particular focus on the dynamics of movement and resource use by animals at the interface between human/livestock and wildlife and the application of geographical information system (GIS) and remote sensing in natural resources assessment and management. She is a Ph.D. candidate at the University of Zimbabwe, co-supervised by CIRAD and supported by the research platform 'Production and Conservation in Partnership'.

Foreword

It was one of the authors of this volume, Professor David Cumming, who with Rowan Martin more than a quarter of a century ago first drew me into the world of serious scholarship on natural resource management. As an ecologist, and at that time the chief ecologist of a government agency responsible for a national parks estate and the stewardship of Zimbabwe's natural resources, his scholarship was centred in conservation ecology. At the same time I was the director of a university department charged with the application of the social sciences to the institutional aspects of rural development, particularly in Zimbabwe's extensive communal lands. The resultant confluence of conservation and development issues was a powerful stimulus to my own scholarship and contributed over time to a new initiative at the University of Zimbabwe conjoining the natural and social sciences in a programme examining ecosystem processes from a multidisciplinary biophysical and social perspective.

Similar developments were taking place in other academic and research centres, tracking international debates on conservation. In a very broad but instructive generalization, it can be suggested that in the 1960s and 1970s the prevailing paradigm of conservation was one of segregation and protectionism, setting aside designated wild places and species from human predation at a time when human activity was becoming more extensive and pervasive. This was the 'fortress conservation' mentioned in this volume. Developed largely during the colonial era in Africa this paradigm suited the political elites of the day. It brought large areas of real estate under the direct control of the state, free of inconvenient indigenous populations. It was achieved at comparatively little opportunity cost, since at the time of their creation parks were generally in areas not in high agricultural demand. The paradigm satisfied the conservation ethos of colonial metropoles and served the recreational needs of the local urban elite.

The creation of national park systems preserved large areas of aesthetically valuable landscape across sub-Saharan Africa, which now form the cornerstone for its tourism industry and an important ecological research base. In most post-colonial states the new politico-economic elite have continued to support these systems as political and economic assets. The problem is that their benefits have not, in general, reached down to the rural small-scale farmers who are the neighbours to these parks. As Chapter 9 of this book comments, these parks and their wildlife became 'a cost to be minimised rather than a benefit to be nurtured'. For these farmers, important political constituencies of governments, parks, wildlife and conservation in general are a low priority when compared to rural agricultural and industrial development. As a result these parks have tended to be underfunded and understaffed and are in some instances degrading. Many survive through the assistance of the international conservation agencies that have traditionally supported 'fortress conservation'.

While 'fortress conservation' was struggling to retain its pre-eminence as a conservation paradigm for Africa in the 1970s and 1980s, a competing paradigm emerged, defining conservation as sustainable use. This was not a new perspective, but it was in these decades that the paradigm saw serious attempts at application in post-colonial Africa. Its essence can be summarized by the following propositions: (a) that the successful stewardship of the rural African environment lay primarily in the hands of those who lived with it; (b) that their investments in the environment should be matched by the benefits they derived from these inputs, and (c), that this 'civil' management of the environment required localized regimes of resource management, legally defined and empowered to take management decisions regarding their natural resources, bearing the cost and enjoying the benefit of such management. This was communal common property proprietorship, or Community-Based Natural Resource Management (CBNRM) as it has frequently been called. CBNRM was not the only mode of management advocated under the sustainable use paradigm; it was also applied to state and private management regimes. However CBNRM, targeted as it was to the vast bulk of the rural African population, was often the paradigm's public face.

CBNRM, with its putative benefits of ecological sensitivity, economic expansion and institutional growth was popular with donors charged with rural development and during the 1990s funding for such initiatives was readily available. Well known are CAMPFIRE in Zimbabwe and similar programmes in other African countries. More broadly, the United Nations Development Conference at Rio and the subsequent Millennium Goals produced at Johannesburg made clear the prevailing view that development and conservation were to be considered as complementary rather than competitive in policy formation.

However, the implementation of CBNRM encountered a number of obstacles, and, with the exception of Namibia, performance fell well below promise. A wide range of factors contributed to this patchy performance and during the first decade of this century scholarship and experience brought more clarity to the obstacles facing successful CBNRM implementation. Among the many factors involved, two can be considered to be fundamental. The first is that the heterogeneity in social and natural resource conditions across rural Africa inevitably means that contexts conducive to communal regimes of resource management are not always to be found. The recent socio-economic history of some areas, with transient and mobile populations, high levels of socio-economic differentiation and poor resource to population ratios, has led to conditions inimical to local collective action. The social capital required for such enterprises is either absent or embryonic, awaiting a restructuring in form and process that has yet to come. Chapters 4 and 5 of this volume provide examples of this situation.

The second impediment to the growth of effective communal governance of the commons lies in the formal, legal tenure status of local CBNRM regimes. At the moment these rights are tenuous at best. They have no clear rights to economically valuable resources. They lack the security required as an inducement to conservation investments. Regulations often preclude the opportunity for experiments in sustainable resource use. They have no negotiating rights and are open to the incursions of agreements made between the government and the private sector. They have no legal persona and are effectively perpetual legal adolescents. As I have said elsewhere (Murphree and Taylor, 2009, p. 109), 'Such conditions remove the incentive for the conjunction of human energy and resource richness . . . in the African landscape and puts in place short-term survivalist strategies which serve neither the interests of populations or the environment.' Unfortunately, genuine devolution involves a transfer of effective rights and responsibilities that governments are reluctant to contemplate, let alone implement. Only in Namibia has legislation pursued this course. Elsewhere innovative use of legislation on trusts or local political pressure under exceptional leadership has provided de facto devolution, but such instances are rare. As a result, initiatives to realize the potential of communal resource management have underperformed. The record has been so disappointing that some have advocated the abandonment of CBNRM approaches. Hutton *et al.* (2005) document this disappointment and the advocacy for a return to a 'fortress conservation' approach in an article entitled 'Back to the barriers? Changing narratives in biodiversity conservation'. This has not happened and CBNRM retains its iconic status, although with less vigour than previously. Conservation and development policy during the first decade of this century has in fact been static, caught between the impossibility of reverting to former and ineffective colonial policies on the one hand and the political centre's resistance to devolving land and resource ownership to the periphery on the other.

Into this static policy vacuum has now stepped the idea of transfrontier conservation areas. The idea has been around for some time, but there is no doubt but that it will be a driving policy preoccupation in the second decade of the twenty-first century. Ironically the 'frontiers' that the idea addresses were set up under colonial dispensations a little more than a century ago and tended to be situated in areas considered marginal for rain-fed agriculture and human settlement. It was not surprising, therefore, that these frontier areas often attracted the designation of nature areas, national parks or state lands. The transfrontier conservation area concept thus has a certain ecosystem logic behind it, although, as this book is careful to point out, one cannot generalize too broadly on this issue. Nor can one overgeneralize on the specific forms that the concept has taken to date. There are now transfrontier parks (TFPs) as well as transfrontier conservation areas (TFCAs), and in respect to the latter the boundaries, status and uses of these areas are frequently undefined, or differently defined. To add to the terminological confusion, these aggregations are sometimes loosely referred to as 'peace parks'.

What are the reasons put forward for the establishment of these transboundary conservation entities? Three categories of motivation are usually put forward. First, that their establishment along national borders will mitigate international antagonisms and promote a spirit of international cooperation on the African continent. Thus the 'peace parks' designation. This I think we must regard as largely palliative rhetoric. International cooperation in Africa is determined by economic and political considerations fashioned at the centre rather than the periphery. If there is border dissension or violence on the periphery, it is carried out by groups to whom the designation of 'peace park' is no deterrent. The phrase is, however, useful in winning the support of politicians, who can put their signature to the enterprise at little or no cost.

A more substantial group of arguments suggests that the transfrontier approach is an antidote to the inefficiencies and absurdities of colonially drawn boundaries which have in many instances cut across ecosystem units and inhibited or destroyed their integrity. Integrated management systems would, it is argued, increase efficiency, bring economies of scale and rationalize ecosystem management demands with transboundary management responses.

A third set of arguments suggests that transfrontier aggregations can overcome the socio-economic marginalization associated with relegation to boundary areas, turning these areas in many instances into new centres of agricultural and industrial growth, associated with transportation and communications networks that cut across previous barriers to commerce and social contact.

While we can largely dismiss the first rationale, we must take the second and third sets of argument seriously. The second set has considerable substance across a range of technical and structural issues. Indeed, if we look at the record of transfrontier initiatives to date we can see that project implementation on the ground is most evident in this sphere. This is not surprising. With the technical and structural research and planning firmly in their hands, the parks' bureaucracies concerned are well placed to plan and implement project aspirations with relative alacrity. Furthermore, the technicist nature of management activities is attractive to the international conservationist community and funding is, relatively speaking, abundant. Nevertheless, integration between two or more management authorities with different resources and styles is bound to be difficult and so even in this set of objectives progress is slow. It is perhaps fair to suggest that many of the transfrontier conservation areas appearing on Figure 1.1 of this volume will never be integrated management units and at best will be planning frameworks for intermittent consultation.

It is on the third set of arguments for the transfrontier initiatives, that which maintains that they can be the engines for socio-economic development in previously marginalized border areas, that we must focus our attention, since on the attainment of this development objective also rests the success of the conservation objectives discussed in the previous paragraph. For the politicians, planners, academics and management personnel who form the epistemic community that usually initiates the substance and implementation of conservation and development programmes this is a critically determinative insight. Their experience and insights are necessary, but not sufficient for success. For transfrontier initiatives to be successful they must become an essential part of the livelihood agenda of those who live within them. They must become an investment of these populations in their own future through a process that links authority and responsibility in clearly defined management regimes, clearly articulated to other management regimes in the conservation area. Without this clarity of definition regarding responsibility, authority and linkages local populations will sabotage the development of the conservation area, either deliberately and openly or covertly and passively.

It is in this arena that the record of transfrontier conservation area initiatives is abysmal. Their establishment has been almost totally devoid of consultation with local populations. The same is true of planning. Structures of communication promoting 'interdependent knowledge production' (cf. Fortmann, 2008, p. 262) and relationships between parks management and the peoples among which they exist are tenuous. In other words, the most important aspect of transfrontier area initiatives has been largely neglected.

At the same time, embryonic efforts to address this failing are in evidence. The transboundary natural resource management fora mentioned in Chapter 8 are a step in the right direction, as is the multi-year local level scenario planning, iterative assessment and adaptive management project funded by IDRC and conducted by the Centre for Applied Social Sciences at the University of Zimbabwe. A further unanticipated offshoot of the Great Limpopo Transfrontier Conservation Area initiative has been the creation of AHEAD (Animal and Human Health for the Environment and Development), started by veterinarians concerned with the veterinary impacts of the initiative. This forum has now expanded to include human health and developmental issues, and has become the premier locus for the multidisciplinary examination of the GLTFCA. It is to be hoped that this multidisciplinarity will be expanded to methodological debates that examine the role of professional researchers and their relationships with local civil science. This third objective of the transfrontier conservation area and its requirements provides an ideal testing ground for what Keeley and Scoones (2003, p. 177) call 'the deliberative approach . . . which places more emphasis on developing methods and institutions that promote communication and address policy issues through inclusive processes of argumentation'.

This foreword has suggested that transfrontier conservation initiatives will feature prominently in conservation and development debates during the second decade of the current century and that the topic has important applied and methodological implications for scholars in the fields of conservation and development. The book provides a penetrating historical review of the shifting geographical, social and political contexts in which these initiatives are found, discusses the socio-economic and ecological factors that impact upon them and, in its final chapters, provides an extensive inventory of the opportunities and the conditions needed to move from opportunities to implementation. If you have an interest in this seminal debate, get the book, read it, keep it handy on your shelf and use it!

Marshall W. Murphree
Professor Emeritus, Centre for Applied Social Sciences
University of Zimbabwe

References

Fortmann, L. (2008) *Participatory Research in Conservation and Rural Livelihoods Doing Science Together*, Wiley-Blackwell, Oxford, p. 262

Hutton, J., Adams, W. and Murombedzi, J. (2005) 'Back to the barriers? Changing narratives in biodiversity conservation', *Forum for Development Studies*, vol. 32, pp. 341–370

Keeley, J. and Scoones, I. (2003) *Understanding Environmental Policy Processes. Cases from Africa*, Earthscan, London, p. 177

Murphree, M.W. and Taylor, R. (2009) 'CBNRM in Africa: current constraints and opportunities', in D. Roe, F. Nelson and C. Sandbrook (eds) *Community Management of Natural Resources in Africa*, IIED, London, p. 109

Acknowledgements

This book is the product of collaboration between the inter-university research programme 'Competing Claims on Natural Resources' of Wageningen University in The Netherlands, the collaborative research platform 'Production and Conservation in Partnership' (RP-PCP) coordinated by the Centre de coopération Internationale en Recherche Agronomique pour le Développement (CIRAD), and the Centre for Applied Social Sciences (CASS) of the University of Zimbabwe. We are very grateful to all collaborators for inputs to the various chapters and to the reviewers who are acknowledged at the end of each chapter. We thank the International Research and Education Fund (INREF) of Wageningen University, CIRAD, the Ambassade de France in Harare, and Canada's International Development Research Centre (IRDC) for financial contributions to the production of the book.

Acronyms and abbreviations

AHEAD	Animal and Human Health for the Environment and Development
BTB	bovine tuberculosis
CA	Communal Area previously known as Native Reserve and TTL (Zimbabwe)
CAMPFIRE	Communal Areas Management Programme for Indigenous Resources
CBD	Convention on Biological Diversity
CBNRM	Community-based natural resource management
CIRAD	Centre de coopération Internationale en Recherche Agronomique pour le Développement
CITES	Convention on International Trade in Endangered Species
CNC	Chief Native Commissioner
DDT	Dichlorodiphenyltrichloroethane (insecticide)
DNPWLM	Department of National Parks and Wild Life Management (Zimbabwe)
DVS	Department of Veterinary Services (Zimbabwe)
EU	European Union
FAO	Food and Agricultural Organization of the United Nations
FMD	Foot and Mouth Disease
GLTFCA	Great Limpopo Transfrontier Conservation Area
GLT(F)P	Great Limpopo Transfrontier Park
GNP	Gonarezhou National Park (Zimbabwe)
IUCN	International Union for Nature Conservation
KAZA	Kavango-Zambezi Transfrontier Conservation Area
KNP	Kruger National Park (South Africa)
LNP	Limpopo National Park (Mozambique)
LSCF	large-scale commercial farming (Zimbabwe)
MZRDP	Mid-Zambezi Rural Development Project
MZRP	Mid-Zambezi Resettlement Programme
NAZ	National Archives of Zimbabwe
NC/DC/DA	Native/District Commissioner (post-1980: District Administrator)
NGO	non-governmental organization

NPA	Native Purchase Area (post-1980 renamed, *see* SSCF)
NPWMA	National Parks and Wildlife Management Authority (Zimbabwe)
OIE	World Organisation for Animal Health
PPF	Peace Parks Foundation (South Africa)
RDC	Rural District Council (Zimbabwe)
SADC	Southern Africa Development Community
SEL	Southeast Lowveld (Zimbabwe)
SNA	Special Native Area (Southern Rhodesia)
SSCF	small-scale commercial farming (previously known as NPA)
SVC	Save Valley Conservancy (Zimbabwe)
TFCA	Transfrontier Conservation Area
TFP	Transfrontier Park
TTL	Tribal Trust Land (earlier known as Native Reserve, now CA)
WWF	World Wide Fund for Nature
ZAWA	Zambian Wildlife Management Authority

1 People at wildlife frontiers in Southern Africa

Jens A. Andersson, Michel de Garine-Wichatitsky, David H.M. Cumming, Vupenyu Dzingirai and Ken E. Giller

> Dream of ancient migration trails trodden deep by an instinct that time has never contained. Dream of a wilderness where the elephant roams and the roar of the lion shatters the night. Dream . . . of experiencing Africa wild and free, where people can reap the benefits of nature and in turn support her. [This dream] will only be realised through the establishment of . . . transfrontier conservation areas.
>
> (Peace Parks Foundation, 2004)[1]

The introduction of peace parks and transfrontier conservation areas[2] (TFCAs) in Southern Africa at the turn of the twenty-first century foretold an enchanting promise: simultaneously contributing to global biodiversity preservation, regional peace and integration, and sustainable economic development. More than a decade later, with TFCA initiatives proliferating (Figure 1.1), the promise remains unfulfilled.

Many people living on the edge of protected wildlife areas now find themselves residing in newly designated TFCAs, but with little clue as to what that actually means. Often displaced by the formation of protected areas decades earlier, they have neither participated in the formation of this new type of conservation area, nor have they started to 'reap the benefits'. With 'conservation' in these transfrontier conservation areas still ill-defined, they continue to live on the wildlife frontier and are exposed to the risks that come with it, while depending on the natural resources of these agro-ecologically marginal environments. Others know all too well what transfrontier conservation means for them. At the core of TFCAs are national parks. To create transfrontier parks or peace parks, new corridors and parks have been defined, effectively putting numerous people 'on the other side'. They now live within protected areas, not only threatened by wildlife but also by displacement. Strikingly, the threat of displacement befalls Mozambicans and Zimbabweans rather than South Africans, in the case of the flagship Great Limpopo Transfrontier Park (GLTFP), which supports the widely held view that transfrontier conservation in Southern Africa is a highly political process dominated by South Africa (see Figure 1.2).

Although establishing transfrontier conservation may be a process requiring decades, the present signs for local people are not good. As conservation and development become linked beyond the national policy arena, we tend to lose sight

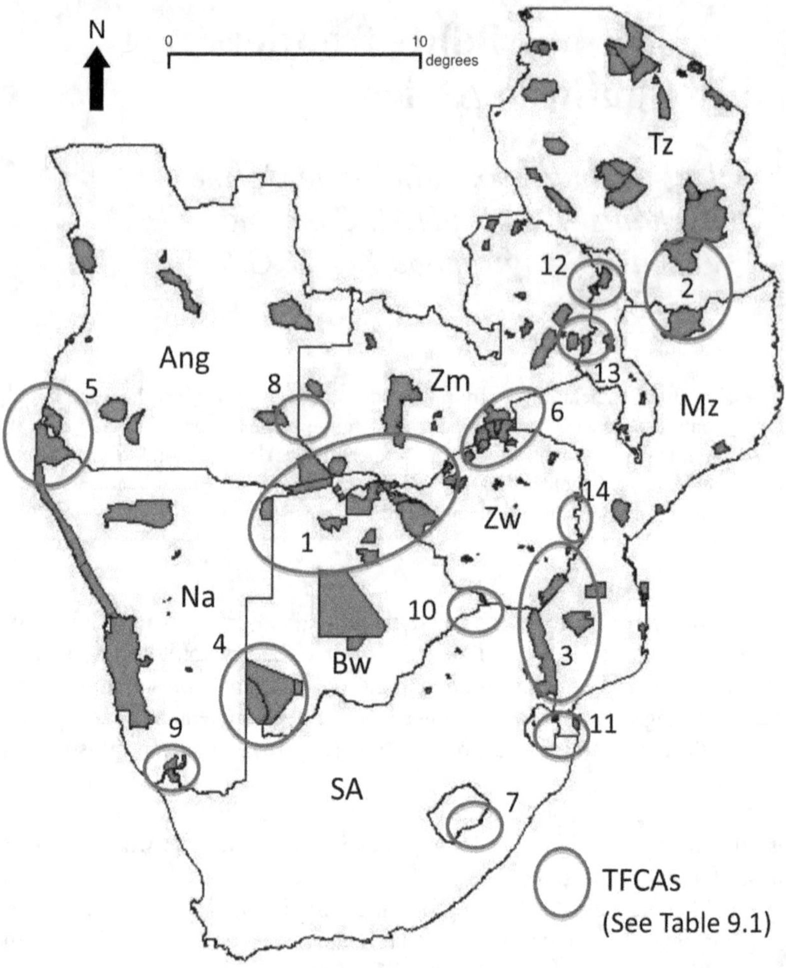

Figure 1.1 TFCAs in Southern Africa (in various states of development).

(1) Kavango-Zambezi; (2) Niassa-Selous; (3) Great Limpopo; (4) Kgalagadi; (5) Iona-Skeleton Coast; (6) Mana Pools-Lower Zambezi; (7) Drakensberg-Maloti; (8) Liuwa Plain; (9) Ai-Ais/Richtersveld; (10) Greater Mapungubwe; (11) Lebombo; (12) Nyika-Vwaza Marsh; (13) Kasungu-Lukusuzi.

of the realities of local peoples living on the wildlife frontier, and the dynamics that drive and affect their resource use. In academic and policy discourse on transfrontier conservation, local peoples may be frequently referred to as 'beneficiaries' or 'victims', but they are not merely marginal in the policy process. They live on the edge in a number of ways. First, making a living on the fringe of protected areas can be a risky business as it is often dependent on the low and erratic rainfall that usually characterizes these areas. Second, there is the – obvious – risk of crop destruction and loss of livestock to wild predators or disease, and even loss of human life to wildlife as the boundaries of protected areas are permeable to various degrees. Third, these agro-ecologically marginal areas, and the permeability of the boundaries, also provide opportunities for both locals and newcomers. Often sparsely populated, these areas are ripe for agricultural expansion. More natural resources can be tapped – often illegally – when crossing the edge. It is on these local realities of people living on the edge of wildlife frontiers that this book focuses. Moving from the dream of eco-tourism-fuelled development and global quests for nature conservation, towards the local realities on the wildlife frontier: what does it mean to live on the edge?

Living on 'the edge'

People in Southern Africa have lived for a long time with wildlife and, for more than a century, within protected areas in which wildlife and other natural resources are preserved.[3] In general, those now finding themselves living on the fringe of protected areas were legally denied access to these lands and their resources – a situation known as 'fortress conservation' (Brockington, 2002). The existence of such an officially enforced divide, however, tells us very little. First, the divide took different forms and changed over time: sometimes merely a line on a map, in other cases a physical barrier – a fence – that restricted the movement of specific wildlife species, domestic animals and humans. In some cases the divide became defined as an intermediate zone – a 'buffer zone' or 'multiple use zone'. These zones were sometimes merely extensions of 'fortress conservation' (Neumann, 1997), in other cases resource access was regulated through concessions or a CBNRM[4] programme. The divide between protected areas and their hinterlands has thus shifted, both in time and space, and in nature. The edge is not, therefore, a zone that can be easily defined in straightforward geographical terms as the area within a defined distance from a protected area. Rather, the edge is defined against different aspects of the protected area it borders, and as a result takes a different form. For instance, when considering risks of disease transmission from wildlife to domestic animals, the edge may be seen as a buffer 'yellow zone', defined by veterinary departments to prevent disease spread from wildlife 'red zone' to disease-free areas of a country, with increased disease surveillance and prevention and restrictions of livestock movements from those 'yellow' areas. But when considering the risk and impact of predation on domestic animals, the edge may take a different form, being a function of both environmental factors within and outside the protected area as well as the predator species concerned.

Second, the meaning of the divide between the protected area and its bordering zones is equivocal and highly dynamic. For instance, the 'fortress' may have been meant to protect natural resources that are within it, but what if the protected go out? Or does the 'fortress' protect those living outside from what is kept within, such as crop-raiding animals, dangerous predators and diseases endangering livestock and people? The divide may have evolved into a difference in resource availability, and therefore acquired meaning as a difference in opportunity. It may be denied altogether, for instance, by those forcibly removed from the protected area. Fences and intermediate zones may also have different meanings: they may be erected as disease barriers or sanitary corridors, to protect livestock and people.

A comparative approach

The captivating notion of transfrontier conservation has not merely resulted in numerous policy initiatives and academic attention. As our understanding of the origins of policy and its implementation increases, we also tend to reiterate the significance of the TFCA concept (Ramutsindela, 2007). A common discourse to talk about and analyse these initiatives has evolved. A first illustration of this, arguably misplaced, reiteration of the significance of the TFCA concept is in the defining of TFCAs and peace parks. Protagonists of TFCAs, and academics alike, increasingly use the terms peace parks and TFCAs interchangeably. This blurs the distinction between the amalgamation of protected areas – (national) parks and game reserves – across national borders into transfrontier parks (TFPs)[5] or peace parks and the larger conservation areas that are often – but not always – constructed around these mega-parks. Although such blurring may suit political ends and fund-raising purposes, the distinction is important as conservation in peace parks (or TFPs) and TFCAs may be of a completely different order. Amalgamated parks across borders (TFPs) usually have elaborate (joint) management structures and clearly defined boundaries and conservation targets. Such TFPs are generally examples of 'fortress conservation', separating (wildlife) conservation within them from human occupation and use of natural resources outside. TFCAs, by contrast, generally include substantial human populations but have no clearly defined formalized boundaries, nor shared conservation rules. A good example of this blurring, and the politics of mapping (Harley, 1990; Hughes, 2001; Taylor, 2008), is provided by maps of the Great Limpopo Transfrontier Park (GLTFP) and the wider conservation area around it, the Great Limpopo Transfrontier Conservation Area (GLTFCA). Map-producing organizations, such as the South African Peace Parks Foundation, not only define the boundaries of the GLTFP differently (or, incorrectly), but also the undefined boundaries of the conservation area, which have been particularly prone to the imperialist imaginations of some conservationists (Figure 1.2).

However defined geographically, the GLTFCA has no joint conservation plan or enforcement of conservation rules. Thus, along the Zimbabwe–Mozambique railway line that cross-cuts the area that is generally perceived to be within this conservation area, one may observe large-scale tree-felling for charcoal production in Mozambique. Another example is that the rural population living in this ill-defined transfrontier conservation area partly survives on wildlife hunting. In Zimbabwe, such hunting may be regulated as safari hunting by tourists, through the Communal Area Management Programme for Indigenous Resources (CAMPFIRE). In Mozambique, it is often unregulated, but not illegal. Similarly, the more recently developed TFCAs that span parts of northern Malawi and Zambia encompass not only national parks (such as the Kasungu, Nyika and Luwangwa

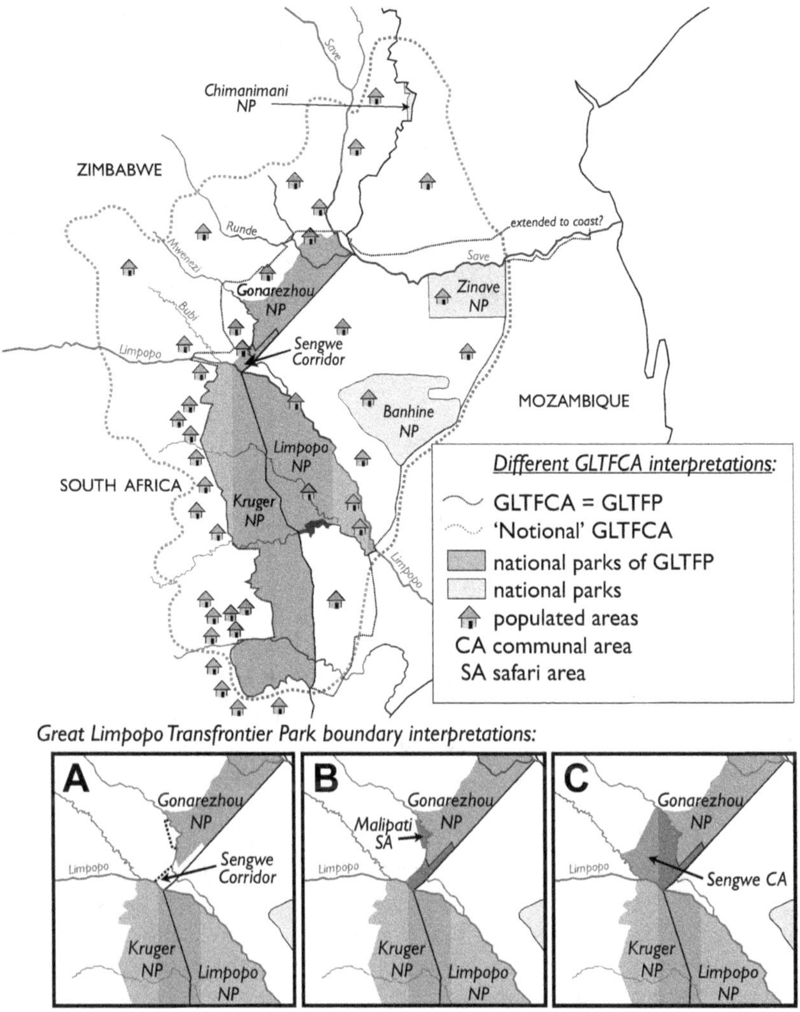

Figure 1.2 Different interpretations of the (potential) Great Limpopo Transfrontier Conservation Area (GLTFCA), and Great Limpopo Transfrontier Park (GLTFP: A, B, C).[6, 7]

(A) UNEP (2008, p. 45).

(B) SANParks and Peace Parks Foundation. As (A), but GLTFP also including the Malipati Safari Area and Sengwe Corridor (approx. 10 km wide) in Zimbabwe. *Sources:* http://maps.ppf.org.za/PDFdownloads%5Cgltp_plain_a3.pdf; www.sanparks.org/conservation/transfrontier/great_limpopo.php (accessed 23 May 2011).

(C) Peace Parks Foundation. As (B), but GLTFP also including Sengwe Communal Area in Zimbabwe. *Source:* http://maps.ppf.org.za/PDFdownloads%5Cgltp_hill_a2.pdf (accessed 23 May 2011).

National Parks) and game reserves in both countries, but also areas of agricultural significance. Located within these TFCAs are major smallholder-farming areas that contribute to deforestation for agriculture and wood-fuelled tobacco curing. And although the TFCA concept suggests synchronized natural resource management practices, what 'conservation' actually means in these areas remains as vague as their boundaries.

While from a perspective strategically oriented to conservation it may be useful to analyse TFCAs as ecological units, it is increasingly recognized that TFCAs are first and foremost political administrative units. To understand 'the edge' and its dynamics requires account to be taken of the diverging situations and trends at the national and sub-national level. It therefore warrants a comparative perspective on 'edges' of TFPs. This book explores such a perspective in the following chapters.

Outline of the book

This book examines what it means to be *living on the edge*. This first entails an exploration of how the people living on the edge of protected areas have been considered (or ignored) in the evolving literature on conservation and development. As will be elaborated in Chapter 2 ('TFCAs and the invisible peoples', Andersson *et al.*), this ranges from invisibility in conservationist perspectives that have propagated TFCAs on (sometimes scientifically questionable) ecological and biological grounds, to that of participants in community-based natural resource management programmes. The creation of TFCAs is seen by some conservationists as a way to restore ecological integrity, and local people are often considered as an impediment to the realization of an effective conservation area. Similarly, the formation of TFCAs in Southern Africa is a highly political process, involving (asymmetrical) negotiations between states represented at the highest level, and figureheads such as Nelson Mandela and the late Prince Bernard of the Netherlands, with the perspective of 'restoring peace' and leading to prosperity and economic development in the region. Yet, local people living on the edge are often absent from political discourses or from academic debates on the 'African renaissance' or the neo-liberal dimension of TFCAs (von Amerom and Büscher, 2005; Dressler and Büscher, 2008). Even the CBNRM literature often misrepresents local people, alternately idealizing them or presenting them as victims in their interactions with the state, international donors, private entrepreneurs or some conservationists who may regard them as 'enemies' of biodiversity conservation.

Chapter 3 ('Defining the edge: boundary formation and TFCAs in Southern Africa' by Andersson and Cumming) explores the meaning and the development of 'the edge' in relation to Southern African TFCAs. People who live adjacent to TFPs are confronted with a complex mosaic of 'soft' and 'hard' boundaries, either as lines on a map, veterinary fences, buffer zones, conservation boundaries, areas of customary authority, overlapping jurisdictions and so on – each with their specific governing regimes. To understand the peoples living in these areas, it is necessary to deconstruct this complex mosaic that constitutes 'the edge'. The chapter focuses on the dynamics of international borders, land tenure, disease management and state-protected area boundaries, the factors influencing their establishment, and how they have changed over the last century.

As the quotation at the beginning of the chapter suggests, local people are expected to benefit from the creation of TFPs and TFCAs. New opportunities in terms of eco-tourism and job creation are often claimed, with some suggesting the existence of a 'honey-pot' effect, attracting people to the edges of protected area edges (e.g. Wittemyer *et al.*, 2008; see also Joppa *et al.*, 2009). Chapter 4 ('Population and livelihoods on the edge' by Giller *et al.*) challenges this perspective on population change. It argues that the drivers of such change are more complex and seldom induced by conservation policies. The chapter describes the heterogeneity of livelihoods on the edge under similar agro-ecological circumstances and analyses the drivers of change in demography and livelihoods. Population changes not only alter pressure on natural resources on the edge, but also result in the coming together of ethnically diverse peoples with differing socio-cultural orientations and livelihood pursuits. Through a number of examples, Giller *et al.* show a huge range in livelihood pursuits and opportunities between locations at the periphery of TFPs in Southern Africa, which lead to a wide variety of challenges for future development of the conservation areas and of local livelihoods.

Southern Africa CBNRM literature has revealed that the assumption of socio-cultural homogeneity of rural 'communities' is problematic. Besides socio-economic drivers, rural communities are also shaped by ideology, ethnicity, history, education and religion. Chapter 5 ('Ethnic heterogeneity and its implications for natural resource management on the edge' by Mukamuri *et al.*) explores such socio-cultural drivers of heterogeneity at the local level, looking at conflicting interests, livelihood trends and worldviews. Mukamuri *et al.* argue that factors such as pre-colonial migrations, colonial forced relocation of Africans, civil wars and recent land reforms are arguably major drivers of 'community' heterogeneity in Southern Africa. Several case studies highlight these differences within planned TFCAs, and the implications for natural resource management.

Paradoxically, the 'edge', often located in marginal agro-ecological areas, is perceived by many as an opportunity for development that is unrelated to its value for nature conservation. Chapter 6 ('On the edge of state and economy' by Dzingirai *et al.*) focuses on this persistent ambiguity of 'the edge'. First, it explores the differing values attributed to it by different actors (entrepreneurs, government agencies, locals, etc). The agro-ecological marginality of these areas, and their assumed 'emptiness', are challenged by the presence of people and entrepreneurial farming on the edge. Simultaneously, the perceived emptiness has given rise to human settlement initiatives. Second, the ambiguity of the edge relates to its proximity to the nation state's borders, and the often weak presence of the state. It is an elusive zone, an arena for a variety of opportunistic and sometimes illegal initiatives. It is also the theatre of opportunistic, and sometimes contra-dictory, interventions by different government departments. Dzingirai *et al.*

conclude that the persistent ambiguity of the edge, shaped by its particular embeddedness in wider developments, makes planned trajectories of mainstreaming and incorporation, including TFCA proposals, prone to deflection.

The 'edge' is thus characterised by heterogeneous communities, with diverse livelihood pursuits. Another prominent characteristic is the uneven distribution of natural resources within TFCAs (DeFries *et al.*, 2007). Chapter 7 ('Resource gradients and movements across the edge of transfrontier parks' by Murwira *et al.*) first examines resource gradients and the ecological processes leading to wildlife movements across the boundaries of protected areas (Battin, 2004). Second, the chapter critically examines to what extent these mechanisms can be extrapolated to livestock and people. Case studies from Zimbabwe and South Africa are used to test the hypothesis that resource gradients drive movements across the edge. They specifically examine the availability of food resources in relation to movements of elephants, prey abundance and movements of wild carnivores, access to grazing and livestock movements, and bushmeat hunting. The chapter concludes that although socio-economic, cultural or political motivations may play a major role, gradients of natural resources influence the movements of wild and domestic animals across the edge. Some possibilities to manage the availability of resources and mitigate the negative impacts of the induced movements are considered.

Indeed, both wild and domestic animals cross the boundaries of protected areas, and Chapter 8 ('Consequences of animals crossing the edges of transfrontier parks' by de Garine-Wichatitsky *et al.*) focuses on the direct and indirect consequences of these movements. Adopting the perspective of people living on the edge of TFPs in Southern Africa, the chapter describes how the proximity of wildlife-rich protected areas generates additional constraints for the people, including physical aggression by wildlife, crop destruction, predation on livestock (e.g. Lamarque *et al.*, 2009), and disease transmission (Osofsky *et al.*, 2005). At the same time it also generates direct benefits, such as additional grazing opportunities for livestock, or revenues and game-meat generated by CBNRM programmes (Taylor, 2009). Recent case studies from Zimbabwe, South Africa and Mozambique, highlight key characteristics of these socio-ecological systems likely to reduce the magnitude of negative human–wildlife–livestock interactions. They also reveal opportunities for positive interactions, which would benefit people within TFCAs.

The ambiguity of the edge also extends to the boundaries of TFCAs, as the areas they encompass are still mostly ill-defined. This has direct consequences for the land and natural resource use options that may generate wealth for those living within TFCAs. Chapter 9 ('Land- and natural resource-based livelihood opportunities in transfrontier conservation areas' by Cumming *et al.*) outlines how opportunities for wildlife-based enterprises vary considerably. Opportunities in TFCAs depend on a country's legal and policy frameworks that regulate natural resource use, population densities, and the social and ecological resources (social and natural capital). First, the chapter examines the context for development and livelihood opportunities in TFCAs. Then it presents some examples of existing opportunities that have been exploited. Finally, the chapter discusses potential options that, given appropriate policies and incentives, can contribute meaningfully to the welfare of those living on the edge.

To conclude, we reflect in Chapter 10 on the need to revisit the objectives of TFCAs and the major challenges these areas face in order to become the locus of conservation *and* development.

Acknowledgements

We wish to thank Aurelie Binot, Harry van der Linde, Kefasi Nyikahadzoi, and Russell Taylor for their comments on earlier drafts of this chapter.

Notes

1 Extract from vision and mission of the Peace Parks Foundation (2004), www.peaceparks.org (accessed 5 January 2010).

2 A TFCA is 'the area or component of a large ecological region that straddles the boundaries of two or more countries, encompassing one or more protected areas as well as multiple resource use areas.' This definition is used in the Southern African Development Community (SADC) Protocol on Wildlife Conservation and Law Enforcement of 1999. The term is similar to IUCN's notion of Trans-Boundary Conservation Areas (TBCAs) (Sandwith *et al.*, 2001; Dudley, 2008). 'Transfrontier' parks (TFPs) or peace parks encompass the protected area(s) that form(s) the core of a TFCA. The concept of 'peace parks', which dates back to the 1930s, stresses the instrumentality of transfrontier conservation in building and maintaining peace (Ali, 2007).

3 The first protected areas in Southern Africa were established in the 1890s.

4 Community-based natural resource management.

5 One of the first UNESCO-recognized TFPs established on the African continent is the 'Réserve de Biosphère Transfrontalière du W' ('Parc du W'), officially established in 2002. It encompasses an area of 17,000 km^2 between Benin, Burkina-Faso and Niger, and rather than a political construct, it was developed through the efforts of the national park managers to jointly address the problem of hunting in parks. A first regional agreement was signed in 1997, and the initiative gradually gained national and international political support. Competing interests with neighbouring (pastoral) communities claiming the right to graze their cattle in the park and to cross international boundaries between the three countries have challenged the sustainability of the TFP. A regional political agreement was reached in 2004. Since then, various stakeholders (including representatives of livestock owners and farmers) have been more formally associated with the management of the edges of 'Parc du W'.

6 The boundaries of the GLTFCA and the GLTFP are often equated. See, for instance: Stalmans and Peel (2010); www.routes.co.za/nationalparks/limpopo/index.html (accessed 23 May 2011); and http://maps.ppf.org.za/PDFdownloads%5Cgltp_plain_a3.pdf (accessed 3 December 2010). The shape and width of the 'Sengwe Corridor', connecting the Gonarezhou and Kruger national parks, varies by source. See, for instance: http://maps.ppf.org.za/PDFdownloads%5Cgltp_plain_a3.pdf (accessed 23 May 2011); and www.cesvi.eu/index.php?pagina=pagina_generica.php&id=745(accessed 3 December 2010). See also Chapter 3.

7 See: Martin (2002). A similar wide interpretation of the GLTFCA's boundaries can be found in: Lindsey *et al.* (2009).

References

Ali, S.H. (2007) *Peace Parks: Conservation and Conflict Resolution*, Westview Press, Boulder, Colorado

Amerom, M. van and Büscher, B. (2005) 'Peace parks in Southern Africa: bringers of an African Renaissance?' *Journal of Modern African Studies*, vol. 43, no. 2, pp. 159–182

Battin, J. (2004) 'When good animals love bad habitats: ecological traps and the conservation of animal populations', *Conservation Biology*, vol. 18, pp. 1482–1491

Brockington, D. (2002) *Fortress Conservation: The Preservation of the Mkomazi Game Reserve*, James Currey, London

Cumming, G.S., Cumming, D.H.M. and Redman, C.L. (2006) 'Scale mismatches in social–ecological systems: causes, consequences, and solutions', *Ecology and Society*, vol. 11, no. 1, p. 14. Available at: www.ecologyandsociety.org/vol11/iss1/art14/

DeFries, R., Hansen, A., Turner, B.L., Reid, R. and Liu, J. (2007) 'Land use change around protected areas: management to balance human needs and ecological function', *Ecological Applications*, vol. 17, no. 4, pp. 1031–1038

Dressler, W. and Büscher, B. (2008) 'Market triumphalism and the CBNRM "crises" at the South African section of the Great Limpopo Transfrontier Park', *Geoforum*, vol. 39, no. 1, pp. 452–465

Dudley, N. (2008) *Guidelines Applying Protected Area Management Categories*, IUCN, Gland

Harley, J. (1990) 'Cartography, ethics and social theory', *Cartographica: The International Journal for Geographic Information and Geovisualization*, vol. 27, no. 2, pp. 1–23.

Hughes, D.M. (2001) 'Cadastral politics: the making of community-based resource management in Zimbabwe and Mozambique', *Development and Change*, vol. 32, pp. 471–768

Joppa, L.N., Loarie, S.R. and Pimm, S.L. (2009) 'On population growth near protected areas', *PLoS ONE*, vol. 4, no. 1, e4279

Lamarque, F., Anderson, J., Fergusson, R., Lagrange, M., Osei-Owusu, Y. and Bakker, L. (2009) *Human–wildlife Conflict. Causes, Consequences and Management Strategies*, FAO, Rome

Lindsey, P., du Toit, R., Pole, A. and Romañach, S. (2009) 'Save Valley Conservancy: a large-scale African experiment in cooperative wildlife management,' in H. Suich and B. Child (eds), *Evolution and Innovation in Wildlife Conservation: Parks and Games Ranches to Transfrontier Conservation Areas*, Earthscan, London

Martin, R.B. (2002) *A Vision for Gonarezhou National Park*, Consultancy for Malilangwe Conservation Trust, Chiredzi, Zimbabwe

Neumann, R. (1997) 'Primitive ideas: protected area buffer zones and the politics of land in Africa', *Development and Change*, vol. 28, no. 3, pp. 559–582

Osofsky, S.A., Cleaveland, S., Karesh, W.B., Kock, M.D., Nyhus, P.J., Starr, L. and Yang, A. (2005) *Conservation and Development Interventions at the Wildlife/Livestock Interface: Implications for Wildlife, Livestock and Human Health*, IUCN, Gland

Ramutsindela, M. (2007) *Transfrontier Conservation in Africa: At the Confluence of Capital, Politics and Nature*, CABI Publishing, Wallingford

Sandwith, T., Shine, C., Hamilton, L. and Sheppard, D. (2001) *Transboundary Protected Areas for Peace and Cooperation*', IUCN, Gland

Stalmans, M. and Peel, M. (2010) 'Plant communities and landscapes of the Parque Nacional de Zinave, Mozambique', *Koedoe – African Protected Area Conservation and Science*, vol. 52, no. 1, pp. 1–11

Taylor, J. (2008) 'Naming the land: San countermapping in Namibia's West Caprivi', *Geoforum*, vol. 39, no. 5, pp. 1766–1775

Taylor, R. (2009) 'Community based natural resource management in Zimbabwe: the experience of CAMPFIRE', *Biodiversity and Conservation*, vol. 18, no. 10, pp. 2563–2583

UNEP (2008) *Africa: Atlas of Our Changing Environment*, Division of Early Warning and Assessment (DEWA) United Nations Environment Programme, Nairobi, Kenya, http://na.unep.net/atlas/AfricaAtlas/AtlasDownload/Maps/great_limpopo.jpg (accessed 23 May 2011)

Wittemyer, G., Elsen, P., Bean, W.T., Burton, A.C.O. and Brashares, J.S. (2008) 'Accelerated human population growth at protected area edges', *Science*, vol. 321, pp. 123–126

2 TFCAs and the invisible peoples

Jens A. Andersson, Vupenyu Dzingirai and David H.M. Cumming

Introduction

Transfrontier conservation represents a combination of differing interests, most notably environmental, political and economic. It is an attractive concept, not only for conservationists and politicians, but also for those concerned with inter-state peace building and regional integration, and issues of economic (under) development (Ramutsindela, 2007; Ali, 2007). The concept is equally popular among scholars of a wide range of academic disciplines, both natural and social sciences. Not surprisingly then, academic interest extends far beyond the confines of ecology, political science and geography. Economists, sociologists and anthropologists have shown a keen interest in transfrontier conservation areas (TFCAs) and their (lack of) developmental impacts, and biologists are interested in how they contribute to conservation of biodiversity. Not only has TFCA formation implications for public and animal health (see Chapter 8), but animal health issues have in turn economic consequences, notably in relation to beef exports to global markets (see Chapter 3). The vast body of academic literature on transfrontier conservation that has emerged over the past decades evidences this interconnectedness of different perspectives; transfrontier conservation appears to be a truly interdisciplinary concept that transcends the natural–social science distinction.

The interdisciplinary character of transfrontier conservation was arguably reinforced by its spread from the north – the first transfrontier park (TFP) was established between the USA and Canada in 1932 (Ali, 2007, p. 2) – to the so-called developing countries in the south. Subsequently, the role TFPs and TFCAs can play in triggering economic development has gained prominence in both academic and policy debates. In Southern Africa, TFCAs have become seen as a vehicle for conservation *and* development.

This chapter explores how people living on the edge of protected areas, and within TFCAs, have been considered in the evolving literature on conservation and development. First, we zoom in on different ecological perspectives that underpin conservationists' arguments to establish TFPs and TFCAs. The central tenet in these perspectives, that larger areas can support a greater biodiversity, provides the ecological basis for conservationists' plans to enlarge and connect conservation areas. It is therefore perhaps not surprising that the establishment of wildlife corridors and protected areas is the most tangible plan pursued within the context of TFCA formation, as both ecological and conservationist perspectives are generally silent on the role and position of people in enlarged conservation areas; they tend to render them invisible.

Ignoring people living in TFCAs may, of course, serve the political and economic interests of those advocating these new conservation areas. TFCA formation in Southern Africa is, after all, a highly political process. Hence, we continue our exploration of the portrayal of people in TFCA thinking and promotion with an exploration of political science perspectives. These have generally focused on understanding the policy success of the TFCA concept, often linking this success to trends such as the 'African renaissance' or neo-liberalism (van Amerom and Büscher 2005; Dressler and Büscher, 2008). In so doing, they too tend to lose sight of the people living within these areas, underwriting merely their marginalization in policy processes.

A third perspective on TFCAs that can be distinguished focuses explicitly on the developmental aspect of conservation and people's involvement in natural resource management. Implicit in much thinking about TFCAs is that people's involvement will be organized as CBNRM. But even the rich CBNRM literature often misrepresents local people, either idealizing them or presenting them as victims in their interactions with the state, international donors, private entrepreneurs or by some conservationists who may regard them as 'enemies' of biodiversity conservation.

Missing peoples: island biogeography, ecological hotspots and wilderness areas

Conservationists generally embrace the concept of establishing transfrontier parks and conservation areas because it is well established that larger areas carry greater numbers of species. They are therefore more likely to support numerically and genetically viable populations than smaller areas. This principle was first established by the pioneering research on island biogeography by MacArthur and Wilson (1967). Their work formed an important foundation for conservation planning (e.g. Soulé and Terborgh, 1999). Areas very rich in species numbers, with a high proportion of endemic species, are referred to as biodiversity hotspots (Myers, 1988; Myers *et al.*, 2000) and provide a basis on which to prioritize conservation action and establish new protected areas (Margules and Pressey, 2000). In Southern Africa, the primary biodiversity hotspots are the Cape floristic region ('fynbos'), the coastal forests of eastern Africa (the Maputaland centre of endemism in Mozambique and KwaZulu-Natal where the Lubombo TFCA is situated) and along the west coast, including the Namibian Desert and the Nama Karoo. The latter areas include the Ai-Ais/Richtersveld TFCA (UNEP, 2008, p. 41).

Preserving endemic species in biodiversity hotspots may, however, not be sufficient to preserve overall biodiversity in a country or region (Bonn *et al.*, 2002).

More recently an additional category for conservation has therefore been formulated, namely, wilderness areas (Mittermeier *et al.*, 1998, 2003). The criteria used by Mittermeier *et al.* (2003) to define a wilderness area were:

(i) that it exceeds 10,000 km^2 in extent;
(ii) it has a population density outside cities of less than 5 people/km^2; and
(iii) that it retains at least 70 per cent of its historical habitat.

In Southern Africa, the Kalahari-Namib is such a wilderness area (including the Kgalagadi and Ai-Ais/Richtersveld TFCAs). Areas with such low population densities seldom occur in any of the other TFCAs of Southern Africa. Other approaches for the identification of priority conservation areas include those dealing with very specific sites such as key conservation areas (Eken *et al.*, 2004) and important bird areas (Barnes, 1998), which often fall within existing national parks and TFCAs.

Besides hotspots and wilderness areas, additional rationales for establishing TFCAs include the belief that linking isolated protected areas across larger landscapes will help to re-establish historical migrations, maintain genetic diversity and also provide opportunities for plant and animal species to more readily adjust to habitat shifts resulting from climate change.

Transfrontier conservation to restore ecological integrity

The notions of ecological integrity, intact ecosystems and ecosystem processes also have a bearing on these rationales in both scientific and public perceptions of TFCAs. Borders may be politically necessary, but they frequently fragment landscapes. These borders may be straight lines that have been drawn on a map with little thought for how they disrupt ecosystem processes. They often consist of a convenient geographical feature, such as a river or a range of mountains, or some traditional trade routes, whereas more appropriate ecological units would include whole river catchments or both sides of the mountains. The common characteristic of Southern Africa's international borders is that they were determined at European conference tables on political grounds with some input from explorers, but ecological and social boundaries were ignored (Chapter 3). Through their creation, conservationists, politicians and the public see TFCAs as helping to restore ecological integrity. By this they do not simply mean the creation of expanded landscapes, although this is an important component of transboundary resource management. They also mean landscapes that share common management systems capable of furthering conservation and development interests across borders.

Tranfrontier conservation to solve specific conservation problems: expanding elephant populations

The establishment of TFPs and linking corridors, together with the removal of fences in some areas, has also been seen as a means of solving conservation problems such as the expanding numbers of elephants in protected areas. It has been suggested that TFCAs and corridors will serve to link presently isolated elephant populations in Southern Africa and create a series of meta-populations that will ease environmental pressures in overpopulated protected areas and provide a basis for natural regulation of elephant numbers (van Aarde and Jackson, 2007). This idea has been particularly prominent in arguments in support of the KAZA TFCA and the burgeoning elephant population of more than 250,000 that is centred on the Chobe and Hwange National Parks in Botswana and Zimbabwe, respectively. However, as Cumming and Jones (2005) point out, elephant populations in Southern Africa are mostly growing at *c.*5 per cent per annum or more and the only land available to absorb increasing numbers of elephants is communal land occupied by smallholder farmers. These rural communities are unlikely to tolerate elephants on their land at densities greater than about one elephant to 5 km^2, and even then only if they receive substantial benefits from having elephants on their land. Under present bans on the trade in elephant products, and restrictions on hunting them, potential benefits, which could be substantial, are unlikely to accrue.

Restoring historical wildlife migration routes

The removal of fences and the creation of an 'Open Africa' or 'Boundless Southern Africa'[1] to satisfy both ecological and social concerns has, in turn, considerable implications for ecological, social and disease dynamics at the interface between protected areas, the surrounding TFCAs, and beyond.[2] Part of the 'Open Africa' dream is to re-establish historical migrations of large mammals, akin to the renowned migrations in east African parks. While it may have been the case in parts of Europe and Asia, there is little evidence to support the claims that large-scale regular migrations occurred in Southern Africa in the past (Cumming, 1999). The only regular migrations found in KAZA, for example, were of wildebeest between Chobe National Park and the Savuti swamp and of zebra within the Makgadigadi system in Botswana, and of wildebeest across the Angola–Zambia border in the Liuwa Plain National Park (Cumming, 2004). Part of the 'Boundless Southern Africa' dream is to re-establish connectivity between populations of large mammals, currently isolated by fences. However, significant movements of animals rarely happen spontaneously and the removal of fences alone will contribute little to the preservation of endangered species, such as some large carnivores and rhinos, which will require additional active measures to ensure the genetic viability of their populations.

The conservation focus for the creation of TFCAs has generally been on Southern Africa's rich large mammal fauna. However, politicians and the Peace Parks Foundation, a leading and politically connected non-governmental organization (NGO) in the development of TFPs and TFCAs, have extended the ecological rationales for the creation of TFCAs to international collaboration, governance of resources at larger scales, regional peace, cultural interchange and economic development. Nevertheless, in their conceptions and imagination of

TFCAs, some conservationists and members of the general public fail to see people in the background. The tendency is to imagine ideal TFCAs as open lands, unsettled and wild.

In this perception of a TFCA, local people are considered to be an impediment to the realization of an effective conservation area and the associated expectation is that they will be resettled elsewhere. The antecedents to this view are the historical evictions that have occurred in the establishment of national parks in the region and the failure to distinguish between TFCAs, which include people and multiple use zones, and the transfrontier parks that lie within TFCAs. Finally there are quiet attempts by the government of Zimbabwe to resettle communities rooted in the country's portion of the TFP (Mombeshora and Le Bel, 2009).

Elsewhere in Southern Africa and in South Africa in particular, there have been major innovative developments in approaches to conservation planning and the extension of conservation beyond park boundaries (e.g. Knight *et al.*, 2006a, 2006b; Rouget *et al.*, 2006). Regrettably these new conceptual developments and practices that fully include people have not yet been deployed in the planning and development of TFCAs in the region (Cumming, 2008).

TFCA formation as political process: power without the people

While conservationists have drawn on ecological thinking to argue for TFPs and linking corridors, conveniently hiding the peoples living in these areas from view, the formation of TFCAs is, obviously, a highly political process. Political scientists in particular have tried to unravel the different interests underpinning TFCA formation, and the, sometimes contradictory, strategies deployed by TFCA protagonists. For instance, while people living in and near designated wildlife conservation areas are often made invisible in the marketing of TFCAs to a global, fund-providing public, TFCA protagonists may stress the economic benefits for local people in regional and (sub-)national political arenas. Typically, its protagonists have presented TFCAs as win–win situations for local peoples' economic development, international peace-building, economic integration and development, and global nature conservation. Hence, understanding why the TFCA policy discourse is so successful is a major theme in political science perspectives on TFCA formation. Yet, as the brief exploration below shows, local peoples' views and interests tend to remain hidden in such analyses.

The success of the TFCA discourse: neo-liberalism as catch-all

One perspective on the political nature of TFCA formation in Southern Africa focuses on the success of TFCAs as a policy discourse. Such understandings relate the growing popularity of the TFCA concept and its rapid expansion in Southern Africa into dominant political and policy discourses and ideologies. Büscher (2010), for instance, views transfrontier conservation as the successor to a hegemonic community-based conservation (CBNRM) discourse in Southern

Africa, driven by a dominant neo-liberal ideology. CBNRM initiatives can be seen as a donor-supported process of decentralization of natural resource management. In transfrontier conservation, it is argued, such decentralization and community-level participation has made way for a global marketing of the TFCA concept to a conservation-interested public and international donors.

Although the global marketing of conservation by powerful organizations such as PPF, WWF and Conservation International is undoubtedly a major factor in understanding the growing popularity of transfrontier conservation, the often-used label 'neo-liberalism' easily turns into an all-embracing, yet obscuring, global force. First, this 'neo-liberal' reductionism ignores a longer trend of global growth in state-protected areas, which dates back to at least the 1970s (West *et al.*, 2006, p. 253). For instance, in southern Rhodesia (now Zimbabwe) and South Africa, both private and state-protected areas developed from the 1950s onwards within the context of a settler-dominated political economy, rather than a free market capitalism (see Chapter 3).

Second, neo-liberal reductionism tends to obscure regional political realities and donor considerations in the redirecting of funds from national CBNRM programmes to regionally focused TFCA programmes in Southern Africa. For instance, the regionalization of donor funding for conservation initiatives in the early 2000s followed national elections labelled 'unfair' in both Zambia (van Donge, 2008) and Zimbabwe (Laakso, 2002; Dorman, 2006). The mounting political crisis in the latter country saw donor support for government programmes dwindle across the board, not merely for conservation. Regionalization of conservation funding is thus likely to have been part of a wider trend in the political economy of development support in Southern Africa.

Third, with the developmental effects of nature tourism in TFCAs still largely 'a dream to be fulfilled', transfrontier conservation still remains a largely state and donor-subsidized enterprise rather than a reality driven by market forces. This is not to deny that markets and marketing have penetrated conservation policy-making. They have for a long time, as exemplified by the conservation policies in colonial Zimbabwe that built on the commercial exploitation of wildlife resources in the 1960s (see Chapter 3). Yet, the influence of market forces in Southern African transfrontier conservation has been limited to particular arenas, notably, fund-raising for conservation by international celebrities and 'ambassadors'. Nature tourism, widely seen as the motor of transfrontier conservation and its benefits, appears less of a market-driven enterprise in many TFCA countries than is often assumed. For instance, operators may complain about lack of free movement for tourists and destination marketing by the TFPs, thus hampering private initiative (O'Leary, 2011).

Fourth, not only the success of TFCAs, but also the popularity of its predecessor in conservation policy making, CBNRM, has been understood as resulting from its complementarity with neo-liberal structural adjustment policies in the 1980s (Nelson, 2010, p. 41). This raises questions regarding the emergence and decline of dominant policy frameworks, the more since the TFCA policy discourse often adopts CBNRM as a component of its conservation and development paradigm.

Neo-liberalism thus tends to become a generalized catch-all in analysing the TFCA discourse, but is of little value in understanding the actual developments, or lack thereof, within and between different TFCA areas in Southern Africa.

The success of the TFCA discourse: anti-politics

Rather than stressing ideological complementarity, a second perspective on the political nature of TFCA formation and promotion stresses the instrumental effects of the depoliticizing TFCA discourse in the understanding of its success. Building on Ferguson's (1994) notion of the 'anti-politics machine', this perspective shows how the active construction of TFCAs as apolitical initiatives serves to legitimize them (Büscher, 2010). By relying on seemingly apolitical notions such as participation, conservation, capacity building and good governance, support is mobilized (Mosse, 2004). Nelson Mandela's statement – which for some time featured on the Peace Parks Foundation's website – on the occasion of the translocation of elephants from Kruger National Park to Mozambique, is illustrative:

> I know of no political movement, no philosophy, no ideology, which does not agree with the peace parks concept as we see it going into fruition today. It is a concept that can be embraced by all.[3]

The TFCA discourse thus works as an 'anti-politics machine', and its successful deployment suspends, or perhaps better, obscures its political effects (Ferguson, 1994: p. xiv). In denying 'politics', the TFCA discourse is, at the same time, instrumental in the expansion of (bureaucratic) power of states and intervening agencies (NGOs).

While political science perspectives on TFCAs go a long way in understanding the success of the TFCA discourse, they tend to be confined to the level of (inter)national policy making and the mobilization of support for TFCAs. Often such analyses remain disconnected from realities on the ground. After all, the successful performance of policy models in national and international fora does not necessarily tell us much about their impact on the ground (compare Rap, 2006). While transfrontier parks have become a reality in Southern Africa, TFCAs continue to be very loosely defined zones around such parks, with no clearly defined borders, nor a codified set of guidelines regarding the conservation of natural resources in these areas. Analyses of the social construction of the TFCA policy discourse rightfully point to its inherent political nature, but often tell us little about its material effects (West *et al.*, 2006, p. 256). People living within TFCAs may thus feature as marginalized by the policy process, but otherwise remain 'flat characters' whose resource use is ill understood.

CBNRM, TFCAs and communities

Since the TFCA paradigm of conservation and development largely builds upon CBNRM as a means to realize development within these areas, it is appropriate to also focus briefly on the ways in which people living on the conservation edge feature in the CBNRM literature. Theoretically, CBNRM is about people, particularly those strategically located in the marginal areas where wildlife lives. In the practical implementation of CBNRM, local people often remain invisible and disengaged. This can be illustrated by looking more closely at its key components, resource governance and benefit sharing.

In respect of resource governance and management, the legal status to manage wildlife resources is frequently bestowed on intermediate state institutions such as rural district councils, which are legal entities created by and accountable to the state (Murombedzi, 1992). It is the councils or their equivalent that enter into partnership with the private sector, such as safari operators, to exploit the natural resources. When they do so, councils claim to be acting on behalf of the local communities, but sometimes take it further and claim that they 'are the people' themselves, and thus represent their own self-interests.

Even where councils' involvement is limited through partial devolution, the authority to manage wildlife is rarely given to community institutions that are downwardly accountable. Occasionally, the partially devolved authority is given to traditional leadership, which may include chiefs or a headman. Ideally, these institutions represent communities, defending them from the state or any other institutions seeking to benefit from natural resources found in communal areas. In practice these institutions are often highly connected to the state whose interests they try to protect (Ribot, 1999). Alternatively, authority may be given to local natural resource committees, e.g. CAMPFIRE committees in Zimbabwe. In theory, such committees represent people. In practice, these tend to represent their own or elite interests, capturing the benefits before they reach the local people. For instance, Nabane (1997) showed how such committees exclude women from management, while Dzingirai (2003) has shown how these committees have, rather dangerously, a blind spot for migrants.

At the level of benefit sharing, local people are not visible either. From the onset of CBNRM, there has been a bureaucratic impulse to concentrate revenue from wildlife at the local government level rather than transmit it to rural peoples who are conserving the wild resources (Murphree, 1993). Often the revenue is retained, against advice from conscientious conservationists and founders of CBNRM, at the council level as management costs in spite of the fact that wildlife management occurs at the local level. Occasionally the balance of wildlife revenue is extended to an unknown number of households, though as in the case of Binga District, Zimbabwe, it remains ambiguous whether this is as a dividend or as actual compensation for frequent but unrecorded costs to rural agriculture (Dzingirai, 2003). Thus, even in CBNRM, local people have not been visible.

Transfrontier conservation seems to maintain the same tradition. To be sure it recognizes and works with the private sector, but it rarely recognizes local people as partners. Only in South African and Botswanan TFCAs is there an attempt to foster community enterprises, but even these are increasingly operated by the private sector on their behalf (Spenceley, 2008; Mbaiwa, 2008).

In terms of participation, which might lead to improved benefits, transfrontier conservation has not done well either. There is no attempt by authorities to work with rural communities. On the Zimbabwean side of the GLTFCA, an Italian NGO tried to facilitate the organization of local people into an association (Dzingirai, 2001). This institution was envisioned to act as a voice for the local people on matters relating to the joint management of the GLTFCA. Beyond this, nothing has been done to ensure that indigenous communities take part in the management and evolution of transfrontier conservation areas. The Zimbabwe–Mozambique–Zambia (ZIMOZA) TFCA initially planned to work with rural people, but after the withdrawal of IUCN, no progress was made in ensuring that communities were part of the evolution of the transfrontier conservation area.

More than in CBNRM, transfrontier conservation does not seem to recognize marginal people, and would prefer them to be invisible. At least, the government of Zimbabwe has been battling to remove the Shangani people from where they are settled so that the image of wilderness is maintained and wildlife moves freely (Mombeshora and Le Bel, 2009). Where people are recognized, it is mainly as people to be discovered and experienced by tourists, or as suppliers of curios and crafts for tourists (Hughes, 2008), rather than as resource users in their own right.

Local people do of course respond to this framing. Sometimes in order to make themselves visible, or to have their voices heard in either TFCA or CBNRM fora, they often resort to subversion, disrupting those conservation or development programmes, which they regard as a threat to their livelihoods. In the Southeast Lowveld of Zimbabwe, for instance, these 'weapons of the weak' (Scott, 1985) included land invasions or illegal settlement (Wels, 2003). Or they may appropriate natural resources illegally as is the custom in Mozambique. This might be through illegal hunting, panning or smuggling, all activities carried out sometimes in non-state places or at night when the exact perpetrators are invisible to the state. Sometimes these desperate activities are carried out alone, but other times in close liaison with distant and invisible outsiders as Duffy (2000) has shown for the ivory trade in Southern Africa.

In the context of CBNRM, and CAMPFIRE in particular, disruptive communities have been described as 'poachers', 'smugglers' or 'panners' and criminalized. In the context of TFCAs, communities reclaiming indigenous territory or exploiting resources in the midst of the GLTFP have been described as 'squatters' in both Mozambique and Zimbabwe. These official labels serve to legitimize the eviction of perpetrators which follows (Dzingirai, 2003). When this happens, communities retreat further or disengage from conservation initiatives to concentrate on their small-scale agriculture or livestock production. Both processes of retreat and disengagement further reinforce their invisibility in conservation and development.

Conclusion

The separation of nature and people in western (academic) thought is arguably a major factor in both the success of the TFCA policy discourse and the invisibility of people living in these areas in that discourse. Although a large range of academic disciplines and interest groups can relate to the TFCA concept, its use and promotion continues to build on the nature–people division. The way in which the TFCA notion is currently used thus raises the questions regarding its interdisciplinary character. Ecological perspectives that form the basis of conservationists' promotion of transfrontier conservation have a blind spot for people. They appear to be outside nature. Not surprisingly perhaps, the impact of TFCA formation in Southern Africa has, until now, largely been limited to the core protected areas, the TFPs. Here, there are no people or, as resettlement efforts and wildlife corridor formation in the name of transfrontier conservation suggests, there should not be people (Milgroom and Spierenburg, 2008).

Clearly there is variation in the way communities within TFCAs are imagined. At one level, TFCAs are presented as wilderness concerns. The people are not factored in the imaginations, and there is no sense that they exist or have interactive relationships with TFCAs. From this perspective, local people are invisible. At another level, TFCAs are seen as zones of interaction; it may be that these people are excluded from their resources, in which case they are victims. Or it may be that these residents are mutually benefiting from TFCAs or CBNRM areas, in which case there are some beneficiaries. Either way, and from this perspective, local people appear to be part of the picture, partially visible.

These imaginations are important because they do inform policies to do with TFCAs. In the chapters that follow we try to characterize people's relationship with TFCAs, hoping also that this might generate or encourage policies that put people at the edge on the centre of conservation and development processes, and we do so using material from the Southeast Lowveld, Zambezi Valley and parts of Hwange in Zimbabwe before doing so regionally in Chapter 9.

Notes

1 See: www.boundlesssa.com.

2 The AHEAD (Animal and Human Health for the Environment and Development) initiative started at the 10th World Parks Congress, held in Durban in South Africa in 2003, exemplifies an attempt to build interdisciplinary links to stimulate research and development policy relating to these dynamics and the ecological, social and economic trade-offs involved, particularly in the Great Limpopo TFCA (Osofsky *et al.*, 2005, Cumming, 2004).

3 Quotation from Nelson Mandela's speech at a ceremony to celebrate the translocation of elephants from Kruger National Park to Mozambique, 12 October 2001: www.peaceparks.org/content/newsroom/news (quoted in Wolmer, 2003: 261).

References

Aarde, R.J. van and Jackson, T.P. (2007) 'Megaparks for metapopulations: addressing the causes of locally high elephant numbers in Southern Africa', *Biological Conservation*, vol. 134, pp. 289–297

Ali, S.H. (2007) *Peace Parks: Conservation and Conflict Resolution*, Westview Press, Boulder, Colorado

Amerom, M. van and Büscher, B. (2005) 'Peace parks in Southern Africa: bringers of an African Renaissance?', *Journal of Modern African Studies*, vol. 43, no. 2, pp. 159–182

Barnes, K.N. (1998) *The Important Bird Areas of Southern Africa*, Bird Life, Pretoria

Bonn, A., Rodrigues, A.S.L. and Gaston, K.J. (2002) 'Threatened and endemic species: are they good indicators of patterns of biodiversity on a national scale?', *Ecology Letters*, vol. 5, pp. 733–741

Büscher, B. (2010) 'Anti-politics as political strategy: neoliberalism and transfrontier conservation in Southern Africa', *Development and Change*, vol. 41, no. 1, pp. 29–51

Cumming, D.H.M. (1999) *Study on the Development of Transboundary Natural Resource Management Areas in Southern Africa – Environmental Context: Natural Resources, Land Use, and Conservation*, Biodiversity Support Program, Washington DC (www.BSPonline.org)

Cumming, D.H.M. (2004) 'Movements and migrations of large mammals in the Four Corners area', in J.R. Timberlake and S.L. Childes (eds), *Biodiversity of the Four Corners Area: Technical Reviews, Volume Two,* Biodiversity Foundation for Africa, Bulawayo, pp. 447–460

Cumming D.H.M. (2008) *Large Scale Conservation Planning and Priorities for the Kavango-Zambezi Transfrontier Conservation Area*, Conservation International, Harare

Cumming, D. and Jones, B. (2005) 'Elephants in Southern Africa: management issues and options', WWF-SARPO Occasional Paper No. 11, WWF Southern Africa Regional Programme Office, Harare, p. 98

Donge, J.K. van (2008) 'The EU Observer Mission to the Zambian Elections 2001: the politics of election monitoring as the construction of narratives', *Commonwealth and Comparative Politics*, vol. 46, no. 3, pp. 296–317

Dorman, S.R. (2006) '"Make sure they count nicely this time": the politics of elections and election observing in Zimbabwe', *Commonwealth and Comparative Politics*, vol. 43, no. 2, pp. 155–177

Dressler, W. and Büscher, B. (2008) 'Market triumphalism and the CBNRM "crises" at the South African section of the Great Limpopo Transfrontier Park', *Geoforum*, vol. 39, no. 1, pp. 452–465

Duffy, R. (2000) *Killing for Conservation: The Politics of Wildlife in Zimbabwe*, James Currey, Oxford

Dzingirai, V. (2001) *Consultation, Planning and Community Interest in Transfrontier Conservation*, IUCN-ROSA, Harare

Dzingirai, V. (2003) '"CAMPFIRE is not for Ndebele migrants": the impact of excluding outsiders from CAMPFIRE in the Zambezi Valley, Zimbabwe', *Journal of Southern African Studies*, vol. 29, no 2, pp. 445–459

Eken, G., Bennun, L., Brooks, T.M., Darwall, W., Fishpool, L.D.C., Foster, M., Knops, D., Langhammer, P., Matiku, P., Radfor, E., Salaman, P., Sechrest, W., Smith, M.L., Spector, S. and Tordoff, A. (2004) 'Key biodiversity areas as site conservation targets', *BioScience*, vol. 54, pp. 1110–1118

Ferguson, J. (1994) *The Anti-Politics Machine: Development, Depoliticization, and Bureaucratic Power in Lesotho*, University of Minnesota Press, Minneapolis

Hughes, D. (2008) 'Third nature: making space and time in the Great Limpopo Conservation Area', *Cultural Anthropology*, vol. 20, no. 2, pp. 157–184

Knight, A.T., Cowling, R.M. and Campbell, B.M. (2006a) 'An operational model for implementing conservation', *Conservation Biology*, vol. 20, pp. 408–419

Knight, A.T., Driver, A., Cowling, R.M., Maze, K., Desmet, P.G., Lombard, A.T., Rouget, M., Botha, M.A., Boshoff, A.F., Castley, J.G., Goodman, P.S., Mackinnon, K., Pierce,

S.M., Sims-Castley, R., Stewart, W.I. and von Hase, A. (2006b) 'Designing systematic conservation assessments that promote effective implementation: best practice from South Africa', *Conservation Biology*, vol. 20, pp. 739–750

Laakso, L. (2002) 'The politics of international election observation: the case of Zimbabwe in 2000', *Journal of Modern African Studies*, vol. 40, no. 3, pp. 437–464

MacArthur, R.H. and Wilson, E.O. (1967) *The Theory of Island Biogeography*, Princeton University Press, Princeton, NJ

Margules, C.R. and Pressey, R.L. (2000) 'Systematic conservation planning', *Nature*, vol. 405, pp. 243–253

Mbaiwa, J.E. (2008) 'The realities of ecotourism development in Botswana', in A. Spenceley (ed), *Responsible Tourism: Critical Issues for Conservation and Development*, Earthscan, London, pp. 205–223

Milgroom, J. and Spierenburg, M. (2008) 'Induced volition: resettlement from the Limpopo National Park, Mozambique', *Journal of Contemporary African Studies*, vol. 26, no. 4, pp. 435–448

Mittermeier, R.A., Mittermeier, C.G., Brooks, T.M.P.J.D., Konstant, W.R., da Fonseca, G.A.B. and Kormos, C. (2003) 'Wilderness and biodiversity conservation', *Proceedings National Academy of Sciences*, vol. 100, pp. 10309–10313

Mittermeier, R.A., Myers, N., Thomsen, J.B., da Fonseca, G.A.B. and Olivieri, S. (1998) 'Biodiversity hotspots and major tropical wilderness areas: approaches to setting conservation priorities', *Conservation Biology*, vol. 12, no. 3, pp. 516–520

Mombeshora, S. and Le Bel, S. (2009) 'Parks–People Conflicts: the Case of Gonarezhou National Park and the Chitsa Community in South-East Zimbabwe', *Biodiversity and Conservation*, vol. 18, pp. 2601–2623

Mosse, D. (2004) 'Is good policy unimplementable? Reflections on the ethnography of aid policy and practice', *Development and Change*, vol. 35, no. 4, pp. 639–671

Murombedzi, J.C. (1992) 'Decentralisation or Recentralisation? Implementing CAMPFIRE in the Omay Communal Lands of the Nyaminyami District of Zimbabwe', Ph.D. thesis, Centre for Applied Social Sciences, University of Zimbabwe, Harare

Murphree, M.W. (1993) *Communities as Resource Management Institutions*, International Institute for Environment and Development, Gatekeeper Series No. 36, IIED, London

Myers, N.R. (1988) 'Threatened biotas: hotspots in tropical forests', *The Environmentalist*, vol. 8, pp. 178–208

Myers, N. R., Mittermeier, R., Mittermeier, C. G., da Fonseca, G. A. B., and Keats J. (2000) 'Biodiversity hotspots for conservation priorities', *Nature*, vol. 403, pp. 853–858

Nabane, N. (1997) 'Gender Dimension in the Communal Areas Management Programme For Indigenous Resources: A Zambezi Valley Case Study', M.Phil thesis, Centre for Applied Social Sciences, University of Zimbabwe, Harare

Nelson, F. (2010) 'Introduction: the politics of natural resource governance in Africa', in F. Nelson (ed), *Community Rights, Conservation and Contested Lands: The Politics of Natural Resource Governance in Africa*, Earthscan, London, pp. 3–31

O'Leary (2011) 'The fine-line between a dream realised and a dream shattered', AHEAD–GLTFCA 11th Working Group Meeting, Kruger NP, South Africa, 2–4 March 2011, www.wcs-ahead.org/gltfca_march2011/day_2/14_oleary_dream_realised-pdf.pdf (accessed 4 May 2011)

Osofsky, S.A., Cleaveland, S., Karesh, W.B., Kock, M.D., Nyhus, P.J., Starr, L. and Yang, A. (2005) *Conservation and Development Interventions at the Wildlife/Livestock Interface: Implications for Wildlife, Livestock and Human Health*, IUCN, Gland, Cambridge

Ramutsindela, M. (2007) *Transfrontier Conservation in Africa: At the Confluence of Capital, Politics and Nature*, CABI Publishing, Wallingford and Cambridge, MA

Rap, E. (2006) 'The success of a policy model: irrigation management transfer in Mexico', *Journal of Development Studies*, vol. 42, no. 8, pp. 1301–1324

Ribot, J. (1999) 'Accountable representation and power in participatory and decentralized environmental management', *Unasylva*, vol. 50, no. 4, pp. 18–22

Rouget, M., Cowling, R.M., Lombard, A.T., Knight, A.T. and Kerley, G.I.H. (2006) 'Designing large-scale conservation corridors for pattern and process', *Conservation Biology*, vol. 20, pp. 549–561

Scott, J. (1985) *Weapons of the Weak: Everyday Forms of Resistance*, Yale University Press, New Haven, CT

Soulé, M.E. and Terborgh, J. (eds) (1999) *Continental Conservation: Scientific Foundations of Regional Reserve Networks*, Island Press, Washington DC

Spenceley, A. (2008) 'Impacts of wildlife tourism on rural livelihoods in Southern Africa', in A. Spenceley (ed) *Responsible Tourism: Critical Issues for Conservation and Development*, Earthscan, London, pp. 159–186

UNEP (2008) *Africa: Atlas of Our Changing Environment*, Division of Early Warning and Assessment (DEWA) United Nations Environment Programme, Nairobi, Kenya, http://na.unep.net/atlas/AfricaAtlas/AtlasDownload/Maps/great_limpopo.jpg (accessed 23 May 2011)

Wels, H. (2003) *Private Wildlife Conservation in Zimbabwe: Joint Ventures and Reciprocity*, Brill Publishers, Leiden

West, P., Igoe, J. and Brockington, D. (2006) 'Parks and peoples: the social impact of protected areas', *Annual Review of Anthropology*, vol. 35, pp. 251–277

Wolmer, W. (2003) *From Wilderness Vision to Farm Invasions: Conservation and Development in Zimbabwe's South-East Lowveld*, James Currey, London

3 Defining the edge

Boundary formation and TFCAs in Southern Africa

Jens A. Andersson and David H.M. Cumming

Introduction

Boundaries are ubiquitous components of the world we live in and exist wherever there is a discontinuity, or asymmetry, within or between objects whether at an atomic, cellular, organismal, local, regional or global level. Boundaries are not confined to the physical, biophysical and geographical spheres. They are also present in behavioural and social interactions and mental states as in the interactions between people that arise as a result of language, religious beliefs, ethnicity and culture, scientific disciplines, class distinctions, access to technology, resource access rights and so on.

In this chapter, we explore the geographical boundaries that exist in TFCAs, the factors influencing their establishment, and how they have changed over the last century. We focus on three TFCAs: the Great Limpopo (GLTFCA), the Zimbabwe components of the Kavango-Zambezi (KAZA) TFCA and the Mana Pools-Lower Zambezi TFCA (Figure 1.1). Boundaries within TFCAs take many forms. They may be clearly demarcated international borders reinforced by security fences and patrolled by armed officials. They may be game or cattle fences, aiming to restrict the movement of disease- and damage-causing animals, or a road or track demarcating different land uses and access rights. Alternatively, they may be merely a line on a map demarcating the boundary of a national park, protected conservation area, district, ward or village, but with no obvious mark on the ground. Such unmarked boundaries may be perceived as transitional zones, such as the risk of encountering dangerous animals. They may be fairly abrupt and narrow or gradual and spreading over hundreds of metres or even several kilometres. The peripheral zone around the core protected areas in TFCAs may thus be characterized by a range of ethnic, cultural and administrative boundaries as well as boundaries relating to the use of resources such as land for cultivation, areas in which livestock may be grazed, and areas where specific kinds of natural resources may be harvested. Human created boundaries in such landscapes are mainly a result of competing claims for resources and changing power relations among the actors and interest groups involved, including state agencies.

We centre on four types of boundaries and their interrelations as well as the drivers and how different scales and levels of governance have influenced their formation: (1) international borders, (2) land tenure boundaries, (3) disease management boundaries, and (4) state-protected area boundaries (Figure 3.1). People living in or adjacent to protected areas, and within TFCAs, have been confronted with a sometimes bewildering mosaic of hard and soft overlapping boundaries of various kinds and varying durations. Boundaries formed have, in turn, often driven the formation of new ones. And where boundaries have remained relatively static, their meaning has often changed, as exemplified by the shift from cattle ranching to game reserves and conservancies on private land next to national parks in South Africa and Zimbabwe.

In the historical discussion that follows, we examine the complex set of drivers of boundary formation within the three TFCAs mentioned earlier, highlighting the dominance of particular drivers in particular historical periods (Figure 3.1). It is shown that boundaries of the core protected areas within current TFCAs were opportunistically gained areas and seldom the result of targeted conservation planning. Like the present-day ill-defined boundaries of TFCAs, protected area boundaries created in the past two centuries were primarily political–administrative constructions, serving the interests of dominant groups in society. They were not only historically specific, but contingent on geographical situations as colonization, and the expansion of capitalist interests was a highly uneven process, not least because of the diverse ecological circumstances within Southern Africa. As the timeline in Figure 3.1 exemplifies, while hunting interests and aesthetic values of wildlife drove game reserve formation in South Africa around 1900, it was land classification boundaries that took shape in Southern Rhodesia at that time. There, the capitalist expansion (settler agriculture) and racial segregation drove land use and land tenure boundaries. The country's first protected areas were formed decades later, when conservationist concerns were mediated by disease control considerations.

We start our exploration of drivers of boundary formation in TFCAs at the larger scale of regional and national levels, discussing drivers of international and protected area boundaries. As we will indicate, such larger scale dynamics also had implications for access to resources by those living on the edge. At smaller scales, boundaries between local jurisdictions (districts, wards, villages), land tenure divisions and cultures also affected resource access and use. Ecological boundaries, notably tsetse infestation boundaries, have affected protected area boundary setting, land tenure divisions and resource use options within TFCAs. In addition, these natural boundaries themselves became the object of intervention, especially in the 1950s, when resettlement and tsetse control hunting and fencing altered the distribution of the tsetse fly. Disease control and land tenure boundary formation thus intertwined, driven by capitalist expansion and the segregationist interests of the then dominant groups in Southern African societies (Figure 3.1).

This chapter presents, in boxes, case histories of the core protected areas (national parks) within the selected Southern African TFCAs to illuminate the complex interplay of drivers of boundary formation. Later chapters zoom in on the characteristics of these TFCA areas (Chapter 4), people's resource use in these

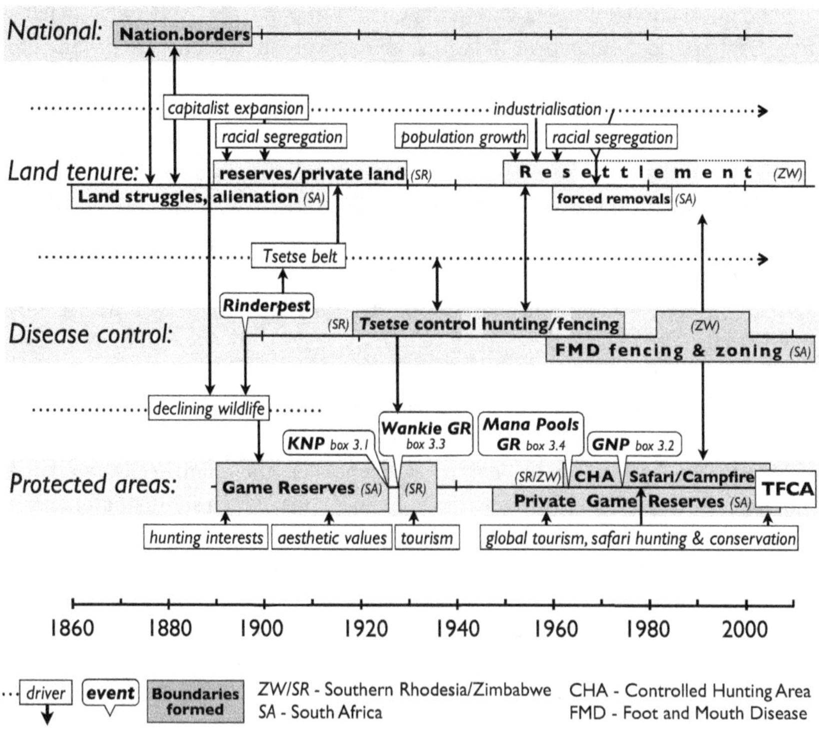

Figure 3.1 Timeline of boundary formation in three Southern African TFCAs. Main types of boundaries, drivers, events, boundaries formed and interconnections.

areas (Chapters 4 and 5), the valuation of these marginal areas on the fringe of the state and national economy (Chapter 6), and the dynamics at protected area boundaries (Chapters 7 and 8). These chapters build on the historical descriptions of the core protected areas presented here.

Crafting national boundaries, 1860–1900

An exploration of the main drivers of boundary formation within Southern African TFCAs logically starts with international borders, as these were established well before any protected areas were created. European settlers and imperial powers crafted them with little if any consideration of African chiefdoms or other local spheres of influence that had been established by those living in these areas (Pakenham, 1991). Such borders were often drawn rather arbitrarily, building on scant knowledge of the continent's interior provided by traders and explorers. Known natural features such as major rivers and watersheds were taken as the initial demarcations between the territories that European statesmen claimed for their countries at the Berlin Conference of 1884–1885, which sought to regulate European colonization and trade in Africa (Wesseling, 1996).

The international borders within the Great Limpopo TFCA, between what are now South Africa, Zimbabwe and Mozambique were, however, settled well before the Berlin Conference. In 1869, a treaty established the boundary between the Afrikaner settler republic and Portuguese East Africa. It stretched from the Limpopo-Luvhuvu River confluence to the boundary between Portuguese East Africa and Swaziland (IBS, 1973; Figure 3.2).[1] The boundary between what was to become the British colony of Southern Rhodesia (now Zimbabwe) and South Africa was agreed upon in 1881 (IBS, 1971a). The boundary between Southern Rhodesia and Portuguese East Africa, which extended from the Zambezi to the Limpopo River, was established after the Berlin Conference, by a treaty signed in 1891 (IBS, 1971b).[2]

The Zambezi River nominally formed the boundary between Northern and Southern Rhodesia (formerly Zambezia under British suzerainty) from 1895, dividing the Tonga peoples living along the Zambezi River. In 1963, when Zambia (formerly Northern Rhodesia) gained independence, the boundary was redefined more precisely, particularly in relation to Lake Kariba, which substantially reduced Tonga territory (Figure 3.5; IBS, 1964).

The border between Bechuanaland (now Botswana) and the Southern Rhodesia border was first delimited by a British Order in Council in October 1898. Joint commissions in 1907 and 1959 defined this border more precisely, and it was marked by beacons and, in places, fences. National boundary making was thus a drawn out process involving imperial and national governments but with only minor adjustments after the initial boundaries were established (IBS, 1974).

Territorial struggles, hunting and early game reserve formation, 1800s–1926

Struggles for territory and political domination were not the prerogative of nineteenth-century European statesmen signing treaties at conference tables. In Southern Africa itself, explorers, missionaries, hunters, prospectors and settlers traversed the interior in search of land and wealth, sometimes also serving imperial or missionary interests. They generally supported their travels by hunting game (e.g. MacKenzie, 1988). The spoils were hides and ivory for export, while the meat obtained along the way served to feed their retinue of porters and guides. As a consequence, wildlife numbers declined rapidly in the nineteenth century, even raising fears of extinction of elephants south of the Zambezi River (Bryden, 1903; see also Cumming, 1981).

Among the various categories of pioneers were Afrikaners seeking independence from British rule in the Cape. From the 1830s onwards, they tried – with varying success – to establish themselves in the northern and eastern Transvaal, negotiating and/or fighting for land with surrounding African polities, such as the Ndebele, Swazi and Pedi (Delius, 1984; Ross, 1997). As with many Africans living in the area, many of the Afrikaners were not farmers, but made their living from hunting and trading in ivory and animal hides (SAHO, undated). Territorial struggles were thus often mixed up with livelihood pursuits and the 'boundary between a hunting and raiding party was tenuous, and often crossed' (Delius, 1984, pp. 33–34).

Hunting traditions and conservationist concerns

As more whites settled in the Transvaal, the exploitation of wildlife resources there increased further, also because sport hunting gained in popularity. Increased competition among the different categories of hunters (Africans, Afrikaners and sportsmen) led to mutual recriminations and attempts to restrict access to wildlife by those who wielded most power. The Afrikaner republic instituted game protective legislation in the mid-nineteenth century to save the valuable commercial resource. But the game laws were not enforced and proved ineffective (Carruthers, 1989, pp. 190–191). Hence, by the 1890s, declining wildlife numbers had become a serious concern for both local and imperial authorities.

Public pressure within the Transvaal, mainly from sportsmen, resulted in the establishment of game reserves on land without agricultural potential or mineral prospects, such as the Pongola Game Reserve (1889) and the Sabi Game Reserve in the South African lowveld in 1898 (Carruthers, 1989; 2007; see Box 3.1, Figure 3.2). Excluding some predators deemed 'vermin', the game reserves initially sought to preserve and increase the number of game animals, 'with the possible future prospect of allowing the excess to be cropped by fee-paying sport hunters (Carruthers, 2007, p. 71). But equally important, the creation of game reserves deliberately sought to exclude Africans from accessing wildlife.

Concerns over the loss of Southern Africa's once abundant wildlife were also expressed by colonial powers, particularly Britain and Germany, both of whom had strong sport hunting traditions. Thus we may understand the inclusion of game laws in the British South Africa Company's (BSAC) Charter for Southern Rhodesia in 1890 (Mutwira, 1989), and the introduction of the 'Convention for the preservation of wild animals, birds and fish in Africa' in London in 1900 (IEA, 2010). The creation of game reserves was also seen as a potential solution to wildlife preservation, their formation being essentially an extension of a long European tradition of royal hunting grounds (Child, 1995), and matched by developments elsewhere. National parks were also being established in North America at the end of the nineteenth century following similar collapses in populations of wild animals through overhunting, e.g. the near extinction of bison.

Capitalist vs. conservationist interests: contested game reserve boundaries and nation-building

While local and global preservationist and conservationist interests played a role in the establishment of Southern African game reserves, the dynamics of when and where these were established and how their boundaries were shaped were more complex. For instance, in the Transvaal, the first game reserves were established in the late nineteenth century (Box 3.1, Figure 3.2), while in Southern Rhodesia the first game reserve was only established in 1928 (Box 3.3). Portuguese East Africa followed later still with human inhabited national reserves and hunting areas (coutadas) in the 1960s.[3]

The early game reserves were also not very stable entities. Some were deproclaimed, such as the South African Pongola Reserve, and competing interests often challenged their boundaries. Developments in the South African lowveld after the Anglo–Boer War of 1898–1902 exemplify this. Land-owning companies, whose ground had been included in the game reserves, now wished to exploit their lands as hunting grounds or for agricultural purposes. The 1896 rinderpest and drought conditions in the 1910s had substantially reduced disease pressure in the Transvaal lowveld, and caused a revaluation of the land as livestock country (Carruthers, 1989, p. 197). But drought also increased 'poaching' by Africans seeking food. Additional pressures on the reserve boundaries came from mining interests and the need to look for land on which to settle Africans, after the passing of the 1913 Land Act that legalized racial segregation in land ownership. Waning wildlife numbers and a shift in the nature of public support for game reserves constituted a countervailing force to such interests. Whereas public support for game reserves had initially been based on local hunting interests, the 1900s saw the emergence of an aesthetic valuation of wildlife by a much wider public as well as increased central government control (Carruthers, 2007, p. 74). The opposing interests of landowners, hunters, conservationists and the Native Affairs Department looking for land for Africans were resolved by land swaps and resettlement (Box 3.1).

Although conservationism drove the expansion of protected areas in the eastern Transvaal, the formation and naming of the Kruger National Park in 1926 was also the outcome of shrewd political manoeuvring of English-speaking and Afrikaner elites. Rising Afrikaner nationalism in the early 1920s witnessed a celebration of the Afrikaner past. Afrikaner politicians deliberately sought to attach an Afrikaner cultural tradition to game protection, despite early Boer settlers' role in the decimation of game in the nineteenth century (Carruthers, 1989, 1994). Thus they garnered support among poor Afrikaners for national park formation. The national park scheme was an attempt to unite English-speaking and Afrikaans-speaking South Africans behind a common ideal (Carruthers, 1994), that is, a conscious attempt at nation-building, but one that excluded Africans. Although Paul Kruger, president of the Afrikaner Transvaal Republic, had supported hunting restrictions in the late 1880s (Grobler, 1996), he had had no active role in the establishment of the Sabi Game Reserve, which later became the southern part of the national park named after him (Carruthers, 1996). As Stevenson-Hamilton, the influential, long-serving warden of the Kruger National Park, once remarked privately:

> I wonder what the old man [Kruger], who *never in his life* thought of wild animals except as biltong . . . would say could he see himself depicted as the *'Saviour of the South African game!!!*
>
> (quoted in Carruthers, 1989, p. 209)

Box 3.1 Africans, Afrikaners and imperialists: the expansion of wildlife conservation domains in the South African lowveld

During the early 1800s, much of what is now Kruger National Park (KNP), was under Swazi control. In 1846, the Afrikaner Commandant Hendrik Potgieter acquired the land between the Crocodile and Olifants Rivers from the Swazis (Pienaar, 1978). The area was not suitable for rain-fed agriculture and was infested with tsetse flies, which limited livestock keeping. Its main use was probably wildlife hunting. The 1896 rinderpest pandemic decimated wildlife numbers. Ineffective hunting legislation made the Afrikaner government proclaim a 'Government Game Reserve' of about 4,600 km^2 between the Crocodile and Sabie Rivers in 1898 (Figure 3.2). After the end of the Anglo–Boer War in 1902, Major Stevenson-Hamilton took on his post of warden of the reserve, a position he was to hold for 44 years. In 1903 the Shingwedzi Reserve (then Singwitsi) of about 8,000 km^2 was established. Stevenson-Hamilton, who was extending the Sabi Reserve westwards (Carruthers, 1996, p. 271), successfully negotiated with land-owning mining companies to protect game on approximately 10,000 km^2 of land north of the Sabie River, thus linking the Sabi and Shingwedzi reserves. The western segment of this area probably included some of the present-day wildlife areas adjacent to the KNP (Figure 3.2).

After the Transvaal became part of the Union of South Africa in 1910, competing economic and conservation interests resulted in a series of land swaps. The area between the Olifants and Letaba Rivers was included in the game reserves but an area along the western boundary was excluded (Pienaar, 1978; Carruthers, 1989). The KNP was gazetted in 1926. While Stevenson-Hamilton had evicted Africans from the original Sabi Game Reserve, removals did not take place further north, as 'it soon came to be appreciated that wildlife protection needed labour and income' (Carruthers, 1996, p. 271). People living within the park were treated as tenants and allowed to remain as long as they did not hunt.

A Foot and Mouth Disease outbreak in 1938 resulted in the slaughter of all livestock in the lowveld. The inhabitants of the park suffered accordingly and most moved out of the, still unfenced, park. Fencing of the western boundary began in 1959 and was driven by disease control requirements. In 1969 the Pafuri Triangle was incorporated in the KNP (Figure 3.2). This involved a swap, with the Department of African Affairs, of an equivalent area of land on the western boundary of the park and the forced movement of the Makuleke people to the area on the west of the park.

With the increasing popularity of KNP in the 1950s much of the land on the western boundary of KNP that was earlier ceded to farmers was converted to private wildlife reserves (Figure 3.2). This trend continued and the fence between KNP and several of its neighbours was removed in the early 1990s (Pollard *et al.*, 2003). Post-apartheid land restitution policy saw the Pafuri area reclaimed by the Makuleke people and it is now managed as contractual park in agreement with SANParks (Robins and van der Waal, 2008). With the formation of the GLTFP, parts of the game and security fence marking the Mozambique border have been taken down to establish ecological links between KNP and the Limpopo National Park.

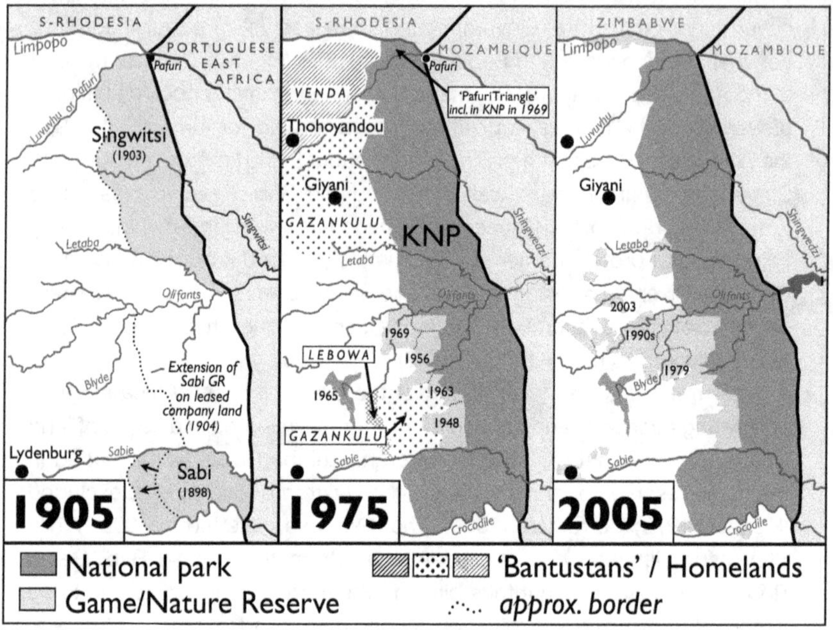

Figure 3.2 Kruger National Park formation in the South African lowveld, 1905–2005. Sources include Pienaar (1978) and Carruthers (1989).

Capitalist expansion, racial segregation and boundary formation, 1890–1930

Rather than preservationist and conservationist concerns, capitalist expansion drove boundary formation north of the Transvaal in the 1890s. The British South Africa Company (BSAC), and white settlers that ventured north of the Limpopo River, became primarily concerned with demarcating agricultural land as the country's mineral wealth appeared to be limited. Well-watered, productive soils in accessible country thus became the markers of land use divisions. While racial segregation soon became an important driver of land use divisions, marking privately owned white settler and communally owned African areas, initially it did not result in large-scale population movements; both the African and settler populations were small. Africans generally stayed on as tenants as there was considerable demand for their labour. The first evictions of Africans from their land only started after 1908, but as in the South African lowveld, Africans living on lands demarcated for private ownership by white settlers usually remained as tenants until the 1940s.

As with the earlier defined national boundaries, internal administrative divisions into provinces and districts paid little attention to African polities and their geographical reach. Natural features, rather than existing political and socio-cultural divisions, were also at this level major markers of the boundaries formed. Thus a layered, and partially overlapping administrative landscape was formed, consisting of colonial state administrative structures and African political structures (chiefdoms and spirit regions). Subsequent subjugating, undercutting, re-inventing and strengthening of customary African chiefdoms – by both government and newly emerging elites – has merely reinforced this dualistic governance pattern, which has often proved problematic (Roe, 1995; Andersson, 1999; Campbell *et al.*, 2001).

Districts became the administrative level at which racial land use divisions and resource use regulations were proposed and implemented. District Administrators (DA), earlier known as Native Commissioners (NC), had considerable influence on the land use divisions within them.[4] Although restricted by land already occupied or reserved for white settlers (alienated land), they were made responsible for the demarcation of Native Reserves, areas for occupation by Africans only. Regarded as a transition until Africans were to become full-time labourers, the Reserves were a direct outcome of the Shona and Matebele risings of 1896 (Palmer, 1977, p. 66). They tried to solve an immediate refugee problem that emerged as a result of the risings and were to 'safeguard' Africans from excessive rent charges by settlers that had taken their land.

Lacking adequate geographic knowledge of the country and clear policy directives, Native Commissioners' boundary setting was foremost a result of their personal assessments of the suitability of particular lands for human occupation and their opinions on the need to cater for future generations (Floyd, 1962, p. 574; Palmer, 1977, p. 58). For instance, around 1900, the NC for Chibi District proposed two large African areas in the Southeast Lowveld stretching to the national border, the Matibi and Chibi Native Reserves. He created such big reserves because water sources were limited and in an attempt to restrict the further expansion of settler lands (Palmer, 1977, p. 254; Figure 3.3). By contrast, the NC for Lomagundi in the northwest proposed 10 small and ill-defined African reserves (Palmer, 1977, p. 259; Figure 3.5).

Although central government sometimes overruled them, and subsequent government committees revised provincial, district and land classification boundaries,[5] NCs remained important advisors and enforcers of the boundaries set. Rivers, soil types, rainfall patterns and accessibility became the major markers of reserve boundaries. Yet, other types of boundaries could have a more direct impact on people's livelihoods. For instance, in the Southeast Lowveld, livelihoods of the Hlengwe people had to a large degree depended on riverbank cultivation, fishing and hunting. Colonial rule meant that fishing was restricted, and cultivation close to a river prohibited (Bannerman, 1978). The BSAC also introduced game laws,

> and in the early years of the twentieth century Africans were regularly stopped from hunting and their guns confiscated – thus criminalising one of the key components of the region's economy and reducing trade in wildlife products.
>
> (Wolmer, 2003, p. 78)

Where relatively clear land tenure classification boundaries had been set, such as in the Southeast Lowveld, Native Reserves were substantially reduced in the first two decades of the twentieth century. The BSAC wanted to make space for large cattle ranches (Box 3.2 and Figure 3.3). By contrast, in areas where boundary-making had been rather rudimentary, as in the western and northwestern part of the colony (Figures 3.4 and 3.5), new and larger Native Reserves were created during the 1910s. But as the NC of Darwin District wrote in 1915, when assigning Native Reserves, 'due attention [was] paid to the exclusion of all such land as will in future be useful and suitable for farming and ranching purposes for European settlers' (quoted in Palmer, 1977, p. 255). The rationale behind the assignment of Native Purchase Areas (NPAs) in the 1920s and Special Native Areas (SNAs) in the 1930s, was no different. While the latter was usually dry, infertile and sparsely populated land set aside as 'safety valve' (see Chapter 6) for later development, the former land category was freehold land for Africans. Meant to create a class of African yeoman farmers, NPAs were often disadvantaged from the start. For instance, the 1925 Land Commission accepted the suggestion of the NC of Darwin District to create a large NPA along the northern border, north of the Mazarabani Reserve (now Muzarabani). The NC defended his suggestion by pointing out that the land was 'useless for European settlers, as it was both far from any railway, waterless, mostly rocks and hills and . . . infested with Tsetse Fly' (Palmer, 1977, p. 255; Figure 3.5). The creation of the Gonakudzingwa NPA, much of which is presently within the Gonarezhou National Park (GNP), provides another example of defining extremely marginal agricultural lands as NPAs (Figure 3.3).[6]

Race-based land tenure boundaries, embodied in the Land Apportionment Act (LAA) of 1930 and subsequent amendments, became increasingly significant after World War II. As will be shown below, with population increases and forced resettlement into the Reserves, these boundaries attained additional meaning as markers of resource gradients.

Box 3.2 Boundary dynamics in Zimbabwe's Southeast Lowveld

The Southeast Lowveld is a meeting place of people of different cultural backgrounds and historically included Tsonga-Hlengwe, Chopio, Sotho, Venda and Nguni (Bannerman, 1978). Around 1750 the Tsonga-Hlengwe people, who were centred east of the Limpopo River in what is presently Mozambique, expanded into what was to become the Southeast Lowveld of Zimbabwe. The subsequent entry into southern Mozambique in the early nineteenth century of the Nguni chief Shoshangane and his creation of the Gaza Kingdom resulted in further movements and changing boundaries of chiefdoms and clans within the Southeast Lowveld. By 1890, most of the Southeast Lowveld was occupied by the Hlengwe, including in what is today the Gonarezhou National Park (GNP).

By 1902 the British South Africa Company (BSAC) had established several Native Reserves in the Colony, of which the Matibi Native Reserve was possibly the largest (Floyd, 1962, p. 573; Figure 3.3). It was not long before the interests of an emergent cattle industry divided the Reserve into Matibi No. 1 and No. 2. The area in between became part of the Nuanetsi Ranch, one of the largest cattle ranches in Africa. The area between Matibi No. 2 and the Portuguese East African border was in 1924 shown as 'Land for Future Determination' and labelled the 'Gona-re-Shaw Forest' (Figure 3.3). In the 1930s, the current areas of the GNP and Sengwe Communal Area appear on maps as a 'game reserve'[7] in which few people lived, but both Europeans and Shangaan hunted (Southern Rhodesia, 1935, 1943). In the 1950s this area was sub-divided into SNAs and the Gonakundzingwa NPA. In the 1960s the NC of Nuanetsi District (now Mwenezi) played an important role in demarcating protected wildlife areas within African areas. He established the Buffalo Bend Game Reserve, the Malipati Game Reserve and Manjinji Pan Bird Sanctuary, the latter two in what became Sengwe Communal Area in 1963 (Figure 3.3; Saunders, 2006, pp. 90–91). The boundaries of the former NPA were revised to a much smaller area upstream along the Nuanetsi River. In 1969 the Gona-re-Zhou Game Reserve was gazetted. This covered an area to the south of the Runde and north of the railway line and included the NC Nuanetsi's Buffalo Bend Reserve. It was only in 1975 that the present boundary and extent of the GNP was gazetted.

The opening of beef export markets in the 1980s, which reinforced the need to control Foot and Mouth Disease (FMD), resulted in new boundaries. A game-proof fence was erected around the GNP, including also sections of adjacent communal areas (Figure 3.7). The 1990s saw few changes to land tenure boundaries but changes in land use. Numerous cattle ranches amalgamated to form wildlife areas, such as the Save Valley and the Malilangwe Conservancies (Figures 3.3 and 3.7). Meanwhile, the different boundaries set have become increasingly contested. In the early 2000s, a chaotic and sometimes violent land reform programme provided the backdrop to opportunistic 'land-invasions' as well as reclaiming of ancestral lands by previously relocated communities.

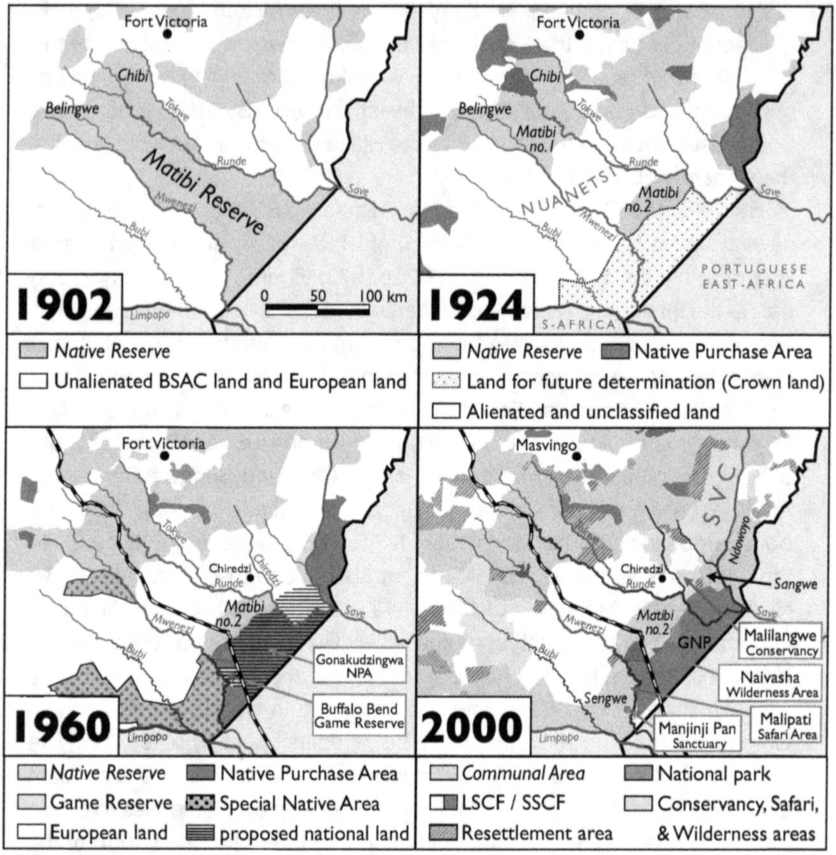

Figure 3.3 Protected Area formation in the Southeast Lowveld, Zimbabwe, 1902–2000 (SVC – Save Valley Conservancy).

Sources: Surveyor General (1924), Floyd (1962, p. 573), Southern Rhodesia (1935, 1943, 1957, 1968), GOZ (2007) and Lindsey *et al.* (2009, p. 177).

Disease threats and state-protected area formation, 1896–1928

Livestock diseases and their management constituted another important driver of boundary formation in the expanding capitalist economy of Southern Africa. South of the Limpopo, colonization and the development of the livestock industry intertwined. Rapid urbanization following the discovery diamonds and gold in the nineteenth century (Meredith, 2008) had driven an increased demand for meat. In order to extend its control over the northern part of the territory, the white settler government sought to develop a 'beef frontier' by settling white farmers specializing on beef cattle production after the end of the Boer War. Yet, the economic viability of the settlement projects in the Transvaal proved problematic. The semi-arid conditions required far larger ranches than the government allocated, land prices were inflated because of expectations of gold deposits, settler farmers were generally undercapitalized and stiff market competition from cheaper African cattle producers reduced investment in the industry. The Transvaal cattle industry thus turned towards highly volatile export markets, and developed on temporary high demand in these markets caused by World Wars I and II (Milton, 1997).

In Southern Rhodesia, the beef cattle industry developed in the wake of company-based colonization by the BSAC. Producing for export to South Africa and the world meat market was capital demanding and, subsequently, it was a business dominated by a few large cold storage companies (Phimister, 1978, p. 398). Despite plenty of apparently suitable land and the availability of cheap breeding stock, the history of the industry is one of undercapitalization, transport problems and disease.[8] As in South Africa, it required continuous state support in order to survive, and the colonial state was heavily involved in disease control from the beginning (Phimister, 1978).

Rather than differences in the epidemiology of diseases such as rinderpest, trypanosomiasis and FMD, dominant societal interests informed disease management strategies and boundary formation. Livelihoods of African cattle keepers were not usually considered whenever such boundaries were set.

Rinderpest and tsetse: state control, resistance and new boundaries

Tsetse infestation of the lowveld areas of the Limpopo and Zambezi river basins had been long recognized by European explorers venturing into Africa's interior (Livingstone, 1857; Selous, 1909; Chorley, 1938). Its epidemiology was relatively well understood in the nineteenth century by both colonizers and resident Africans (Brown, 2005, p. 525). The presence of trypanosomiasis has been regarded a major factor limiting the advancement north of ox-cart dependent Afrikaners escaping British rule of the Cape Colonies (Chorley, 1938, p. 5). Yet, in the 1870s, the tsetse fly belt in the Limpopo Valley had started to break up, enabling cattle-drawn wagons to pass north more easily, through the British protectorate of Bechuanaland, present-day Botswana (Chorley, 1938; Jack, 1933, p. 366). But a cattle-based transport system and the importation of cattle in the early colonial period were to prove disastrous for the emerging Rhodesian livestock industry in the early 1900s.

In 1896 a rinderpest pandemic swept through Southern Africa. It decimated cattle herds and certain wildlife species (notably buffalo and wildebeest). The colony's ox-based transport system ground to a halt (Mutowo, 2001). The consequences of the pandemic were far-reaching, impacting also boundary formation in future TFCAs. First, the BSAC took preventative measures, such as the culling of African herds. These measures aggrieved African cattle keepers and contributed to the Matabele and Shona risings of 1896 (Ndlovu-Gatsheni, 2009, p. 49). The risings, in turn, made the provision of land for Africans an acute one for the colonizers, and thus constituted an important driver of the racial division of land (Palmer, 1977; Phimister, 1988, p. 68) and the associated boundaries.

A second result of the rinderpest pandemic was its effect on the tsetse infestation boundary.[9] Tsetse flies, the vector of trypanosomiasis, lost their host animals (wildlife and livestock) in the pandemic, and the disease disappeared from the lowveld of the Transvaal. In Southern Rhodesia tsetse fly had effectively disappeared from the Southeast Lowveld (Selous, 1909; Lovemore, 1994). Consequently, the area became regarded as excellent cattle country, and hence a target for land alienation (see above). In the northern part of the country the 'fly area shrunk to few comparatively small patches in the districts of Sebungwe, Hartley and Lomagundi' (Jack, 1933, p. 367; Figure 3.5) but as the natural host populations of the tsetse fly recovered relatively rapidly, such as in the Sebungwe District in the Zambezi River basin, re-infestation followed suit (Lovemore, 1994). Pockets where tsetse had survived became reconnected again, resulting in a 'vast fly area extending from the Wankie District in the west, to Darwin District in the northeast' (Chorley, 1954, quoted in Lovemore, 1994; see Figures 3.4 and 3.5). Tsetse presence thus was an important driver of land tenure boundaries, resulting in the designation of such lands as (future) African areas (Jennings and Huggins, 1935, p. 302).

Tsetse control and wildlife conservation: setting game reserve boundaries

The 'fly belt' not only affected the racial land tenure classification at the end of the nineteenth century. It had other implications for boundary formation as well. As the link between ungulate presence and tsetse occurrence was established, tsetse control policy became based on eliminating the fly's host populations of wild ungulates.[10] Thus, when tsetse infestation threatened settlers' mining or agricultural interests in the Hartley District in 1905, free shooting of wildlife was instituted. And although the fly only completely disappeared after parts of the area were deforested to provide wood for a local mine in 1912 (Jack, 1933, p. 368), game elimination nevertheless became the dominant means of tsetse control.[11]

A game elimination programme started in 1918–1919 to halt the spread of tsetse in the vicinity of the Gwaai–Shangani River confluence (Figure 3.4). Tsetse control hunting gave rise to a new set of boundaries, as it was aided by extensive game fences and game corridors (Jack, 1933, p. 369; Cumming, 1981; Taylor and Martin, 1987, pp. 328–9). The game elimination effort, led by the government's Chief Entomologist, was also assisted by (re)settlement policies and forced destocking in affected Native Reserves. For instance, a scheme to attract a 'hunting class of settler' to the Gwaai River area was instituted in 1928, while African hunters were armed in the Shangani, Sipolilo (now Guruve) and Magondi Reserves in the 1920s. Government also removed all cattle from the Magondi Reserve in 1929, and again from the Urungwe Reserve (now Hurungwe Communal Area) in 1953 (Jack, 1933; Southern Rhodesia, 1953, p. 101; Figure 3.5), thus seriously affecting livestock dependent smallholder agriculture in these areas. By 1930, it was realized that cordoning off the expanding fly area was probably all that could be achieved (Jack, 1933).

The shifting fly belt and Southern Rhodesia's tsetse control policy also affected the government's position towards protected wildlife areas. First, a late 1920s plan to establish a transfrontier protected area along the borders of South Africa, Southern Rhodesia and Portuguese East Africa – covering an area similar to the current GLTFCA – was thwarted by the Rhodesians and Portuguese fearing the spread of diseases, tsetse and livestock predation (Mavhunga and Spierenburg, 2007, p. 118). Second, the tsetse issue affected the location of protected wildlife areas and their boundaries within Southern Rhodesia. As the government's Chief Entomologist explained the location of the first large game reserve[12] in the Wankie District in 1928,

> The establishment of game reserves in country infested with tsetse fly has . . . been studiously avoided. . . . no really large area of land suitable for a game reserve was available outside the limits of the former occurrence of the tsetse fly, but there appears a 'reasonable prospect' of keeping the fly from re-invading the main game reserve in the Wankie district.
>
> (Jack, 1930, p. 831)

Thus we may understand why the country's first game reserve was not established in the relatively game-rich and sparsely populated Sebungwe region of the country. Instead, a largely waterless area in the Wankie District with little game in it became the country's first game reserve in 1928. It deliberately excluded the area near the Gwaai River (now Gwayi) where, as the first warden of the Reserve noted, game tended to concentrate in the dry season (Davison, 1967, p. 7). In this area game had been eliminated in the first tsetse programme in 1918–1919 (see above; Child and Riney, 1987), while in the Wankie Game Reserve water supplies had to be developed for wildlife numbers to be able to increase, and thus, attract tourists. Later extensions of the reserve boundary to the perennial Deka River (Box 3.3, Figure 3.4) improved the natural water situation of the Reserve, while still adhering to Chief Entomologist's dictum quoted above.

The proclamation of Wankie Game Reserve in 1928 (see Box 3.3) marks the beginning of a period of protected area formation in Southern Rhodesia. In 1929, a 'Game and Fish Preservation Act' was passed enabling the proclamation of

Box 3.3 Shifting boundaries in western Zimbabwe, 1894–2000

Kalahari Sands cover the greater part of western Zimbabwe. The area is waterless for much of the year and is poorly suited to agriculture. The San people had long occupied the area. Following colonial occupation in 1890, the first demarcation of boundaries occurred in 1894 when two Native Reserves were established, the Gwaai and Shangani Reserves (Figure 3.4; Floyd, 1962), to accommodate the defeated Ndebele under King Lobengula. The reserves were already home to small ethnic groups, including the Tonga, the Nambiya, Dombe and the San. Apart from some land being taken by European settlers near the Victoria Falls there was little settler penetration of the area before 1925 (Palmer, 1977). However, the area along the Bulawayo to Wankie railway line, which was completed in 1903, had been indiscriminately logged for its Zambezi teak (*Baikiaea plurijuga*), much of which was used as sleepers in the construction of the railway.

By 1920 the Wankie Native Reserve was demarcated using the railway line as a boundary marker. By 1924 the Gwaai Reserve was split into the Gwaai and Nata Reserves, NPAs and the Gwaai and Ngamo Forest Area were formed. The remainder of the area to the west of the Gwaai and north of the Nata Rivers was mapped as 'Land for Future Determination', effectively unassigned crown land (Figure 3.4).

Wankie Game Reserve was gazetted in 1928 and not long afterwards Robins donated his farm and private game reserve along the Deka River to the state and so extended Wankie Game Reserve (Davison, 1967). No fences demarcating land use areas were erected. These were to come in the 1970s when fences were erected to control game and cattle movements in an effort to control FMD along the boundaries of what are now Hwange National Park and the Matetsi Safari Area.

During the 1950s the Victoria Falls, Kazuma Pan and Wankie National Parks were gazetted and an SNA was designated to the latter's south; an area occupied by the San people. In 1963 a Controlled Hunting Area (CHA) was established on the northern and eastern boundary of Wankie National Park, and additional forest areas (Sikumi, Fuller and Panda-Masui) were gazetted. The CHA was later incorporated into Hwange National Park and into the Deka Safari Area. The Nambiya and Dombe people were pushed further east by these designations (Figure 3.4).

The European commercial farming area north of Wankie National Park was appropriated by government in 1972 for use as wildlife area and redesignated as a Safari Area in the Parks and Wildlife estate in 1975. Increasingly, wildlife utilization was being seen as the most viable form of land use in the region and by the early 1980s most of the land, both state and private land, north of the Gwaai Reserve (Tsholotsho Communal Area) and even to the east of the railway line was under wildlife. A final extension was to occur with the advent of CAMPFIRE when hunting concessions were established along the southern boundaries of Hwange National Park in the Tsholotsho Communal Area.

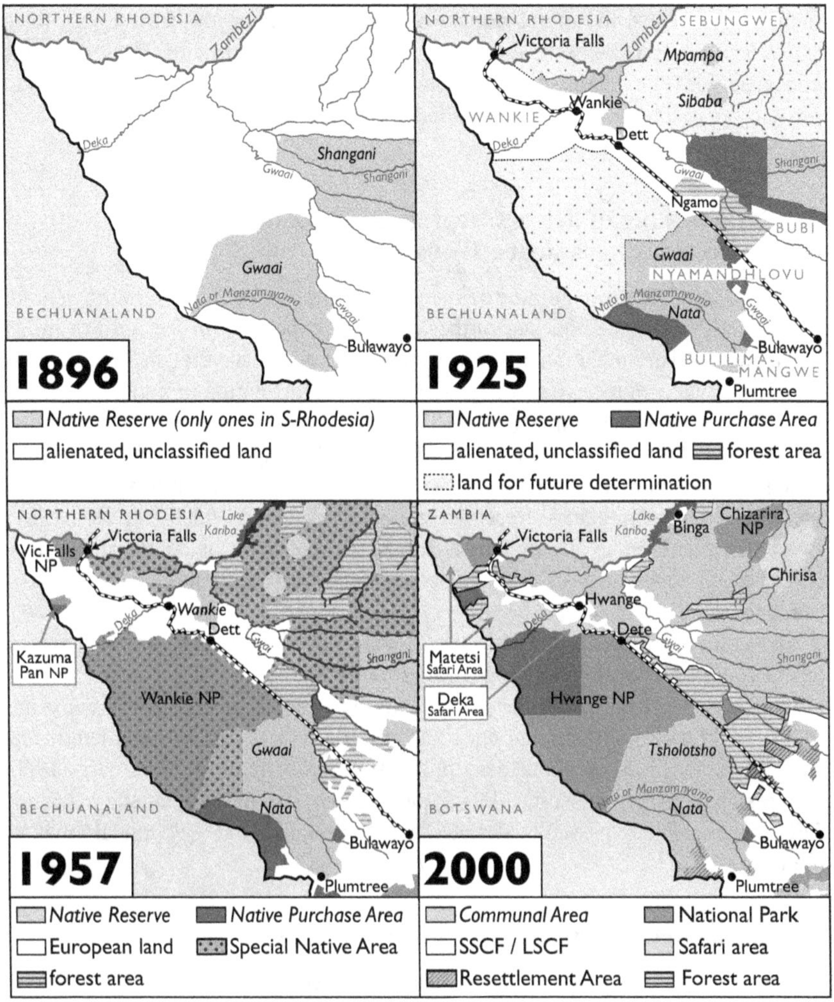

Figure 3.4 Shifting boundaries in western Zimbabwe.

Sources: Surveyor General (1924), Southern Rhodesia (1935, 1943, 1957), Floyd (1962), GOZ (2007).

non-hunting game reserves (Child and Chitsike, 2000; Child, G., 2009, p. 70). Southern Rhodesia also signed an international convention in London in 1933, undertaking to establish national parks (Ranger, 1999, p. 137).[13] By the mid-1930s Victoria Falls, Wankie and the Kazuma Pan Game Reserves were dedicated to game preservation (Mutwira, 1989), as was the privately owned Robins Farm (Box 3.3).

Also in the Southeast Lowveld a game reserve was apparently established in the unassigned area (Box 3.2 and Figure 3.3). The advance of tsetse infestation in Portuguese East Africa and on the border with that country in the 1940s and 1950s (Ford, 1960) may well have influenced its redesignation as an SNA and NPA and later, in the 1960s, the establishment of a CHA to the north of the Lundi River (now Runde) (see below, Figure 3.6). Balancing the growing conservationist concerns with the established interests of the settlers' livestock industry and a growing population became the challenge of state-driven boundary formation in the post-World War II period.

The intertwining of drivers: segregation, tsetse control and induced resource gradients, 1940s–1960s

The 1940s marked the beginning of an upsurge of conservationist thinking in South Africa and Southern Rhodesia, further capitalist expansion in the form of industrialization and urbanization, as well as massive state intervention in African areas. Concerns over progressive soil degradation and deforestation turned the notion of 'carrying capacity' into a watchword in government policy making (Southern Rhodesia, 1939; Bolding *et al.*, 2003). In South Africa, the so-called 'betterment schemes' were already outlined in the 1936 Land Act, but large-scale government intervention in Native Reserves started after World War II (de Wet, 1989). Land use planning in South Africa forced people to relocate to, and within, Reserves, while forced destocking, limits on land holdings and 'modern' agricultural practices were imposed.

In Southern Rhodesia, similar policies were enforced in many Native Reserves in the 1940s and 1950s. Interventions first targeted the 'overpopulated' Reserves with severe land degradation. But as implementation progressed slowly and increasingly met with African opposition, numerous less densely populated areas had not yet been done when these policies were abandoned in the early 1960s (Phimister, 1993; Andersson, 1999). Thus, sparsely populated areas such as future TFCAs were often bypassed by destocking measures and land use boundary making (centralization).[14]

Population growth, carrying capacity and forced resettlement

The notion of carrying capacity, which underpinned the interventions in African land use, also informed forced population movements into the reserves in the 1940s and 1950s. However faulty its 1940s assessment may have been, carrying capacity figures were used to determine where people from reserves considered 'overstocked and overpopulated' needed to go (Southern Rhodesia, 1946–1956). As a result, 'underpopulated' reserves on 'the edge', and in agriculturally marginal areas, became the recipients of immigrants from 'overpopulated' areas. Africans were also moved from European lands were they had hitherto been allowed to stay as tenants, and from Crown lands that had been newly designated as commercial farmland to absorb the post-World War II wave of European settlers.

Although the impact of such interventions in African areas on future 'conservation edge' areas was often limited at the time, they did entail a stricter enforcement of earlier established land tenure boundaries. This eventually resulted in high human population densities on the edge, increasing pressure on natural resources and the formation of hard edges between protected areas and communal lands. The Matibi No. 2 Communal Area, adjacent to Gonarezhou National Park (Figure 3.3) provides an example of the consequences of these earlier resettlement policies and population growth. The human population of Matibi No. 2 grew from about 2,500 in 1920 to 50,000 in the year 2000 while the area available per household declined from *c.*800 ha to 35 ha per household. To make a sustainable living from the land in this area, it is estimated a household requires several hundreds of hectares (Cumming *et al.*, 2006).

Tsetse control and segregationist policies: setting new boundaries

In the 1950s, tsetse fly and trypanosomiasis control through indiscriminate game elimination was gradually replaced by other measures. These resulted in new boundaries in areas that have recently become part of TFCAs. Game elimination continued, but became more specific, targeting the main host species only. Selective game elimination became increasingly confined to CHAs and corridors (Cumming, 1981, 1989). For instance, a 600-mile long cordon of some 10–20 miles wide along the edge of the fly area in the Zambezi River basin was established in the 1960s (Cumming, 1981). Extensive bush clearing and the resettlement of Africans into tsetse-infested areas were additional strategies to shift the tsetse infestation boundary.

Tsetse control and segregationist policies became mutually reinforcing in shifting existing boundaries as more and more Africans were evicted from European lands into Native Reserves in the 1940s and 1950s. As many Native Reserves had become densely populated – 'overpopulated' in colonial policy discourse – relatively sparsely inhabited ones, SNAs, and hitherto unassigned lands became targets for resettlement (Southern Rhodesia, 1946–1956). Thus many Africans moved, or were moved, into tsetse-infested Native Reserves and SNAs in the present-day Binga, Kariba and Gokwe districts (Cumming and Lynam, 1997; Chimhowu, 2002, p. 555; Nyambara, 2001; Dzingirai, 2003).[15] In the early 1950s, 'game destruction, aerial spraying, traffic control and the removal of cattle [had] contributed to the reduction' of tsetse in these areas, but also resettlement was used as a deliberate tsetse control strategy as,

> Human settlement is believed to have effects on the fly population. Although man is a possible source of food for fly, it is reckoned that his activities in clearing and cultivating land . . . are likely to upset the balance of the fly's

Box 3.4 Shifting boundaries in the Zambezi Valley, Zimbabwe, 1924–2000

In recent historical times the Tonga people occupied the Zambezi Valley between Zimbabwe and Zambia. On the southern side of the Zambezi River, the Tonga occupied the present-day Binga and Kariba districts while the area between Kariba and the Musengezi River was occupied by the Korekore and Doma peoples. The area surrounding Feira had a long history of contested boundaries involving early Portuguese invasions and resident peoples such as the Doma. Following colonial occupation, there was little penetration of the Zambezi Valley by European interests, as the low-lying area beyond the escarpment was tsetse infested (Jennings and Huggins, 1935, p. 302). A first designation of Native Reserves in 1902 proposed several small but undefined reserves. By 1915, the Mazarabani Reserve was designated. The remaining area from the Musengezi River westwards to the Shangani River was still mapped in 1924 as land for future determination (Figure 3.5). The area was within the fly belt and was considered unsuitable for habitation and agriculture and thus it remained until the 1950s. With the construction of the Kariba dam in the 1950s the Tonga living along the river were forced to move inland and were resettled in various areas of the Sebungwe, an area comprising present-day Binga, Gokwe and Nyaminyami districts. To the east, the Dande SNA was established. The intervening area between the Dande and westwards to the Rukomechi River (then Rekometje) was designated as 'proposed new national land' in 1960.

By the early 1960s, the demand for recreational hunting by local and overseas hunters had grown considerably. As a consequence unpopulated and sparsely settled areas in the valley such as Urungwe, Rukomechi, and Chirundu became CHAs. In 1963 the Sapi CHA was added and Mana Pools and the Chewore areas were declared non-hunting reserves. In 1968 the Dande and Ruya Game Reserves were established, which were intended to provide benefits from wildlife to local communities.

As in the Southeast Lowveld, the control of tsetse fly and accompanying game fences and corridors in which selected game animals were eliminated played a role in establishing boundaries along the Zambezi escarpment. A game fence demarcated the southern boundaries of the parks and wildlife areas, from the shore of Lake Kariba to beyond Mushumbi Pools. A part of the Mukwichi Communal Land was fenced within the wildlife area adjoining the Chewore Safari Area.

Over the next seven years, the wildlife areas in the Zambezi Valley were expanded and finally consolidated under the 1975 Parks and Wildlife Act into their present boundaries (Figure 3.5).

life. Many species of game would also be driven away . . . it may be that settlement drives out numbers of smaller animals, which are not reached by a shooting policy.[16]

Hence, from the 1950s onwards, future TFCAs became increasingly settled, even when tsetse infested (Nyambara, 2001). The effectiveness of game elimination and spraying against tsetse became largely a function of government investment. When effort temporarily declined, such as during World War II, a brief termination of game elimination in the early 1960s and the late 1970s liberation struggle, this had direct consequences on the tsetse boundary (Lovemore, 1994), resulting in re-infestations (Chimhowu, 2002). And although the total African area increased (Murphree and Cumming, 1993, p. 141), the rapidly growing population in these areas increased pressures on natural resources, making administrative land tenure boundaries on maps more and more visible in a still largely unfenced landscape. Increasingly abrupt resource gradients, in turn, have given rise to exchanges (such as grazing arrangements) and contestations over boundaries.

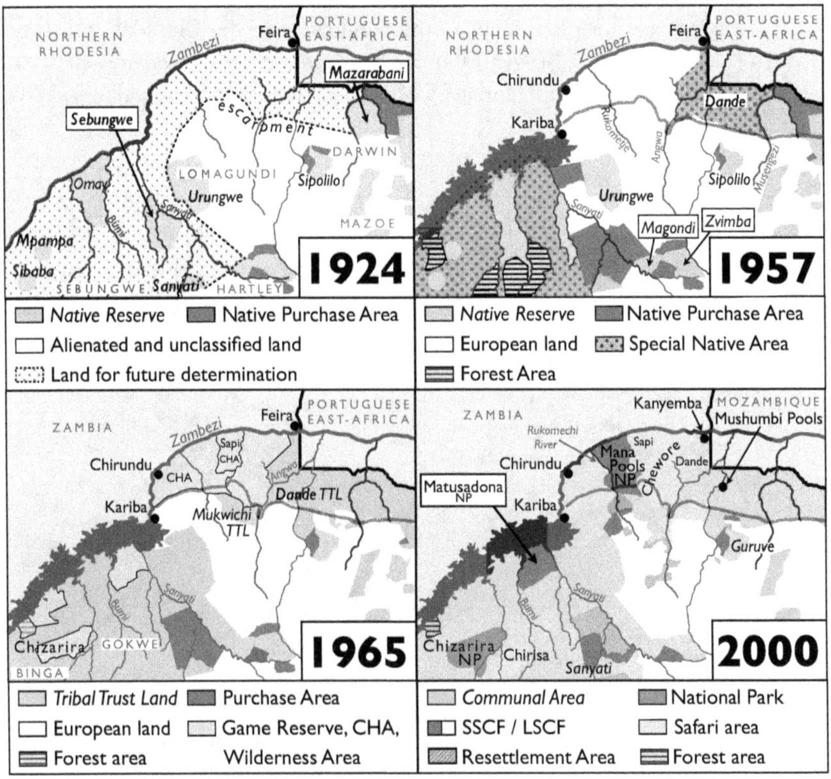

Figure 3.5 Shifting boundaries in the Zambezi Valley.

Sources: Surveyor General (1924), DWC (1962), DNPWLM (1963–1975), Southern Rhodesia (1943, 1957, 1968), GOZ (2007).

Tsetse in the Southeast Lowveld: new boundaries restricting resource access

The changing tsetse situation in the Southeast Lowveld provides an illuminating example of the effect of the government's control policies on boundary formation and rural livelihoods. As in the northwestern part of the country, tsetse reappeared in the lowveld after the 1896 rinderpest pandemic, spreading from Mozambique. By the 1930s the Rhodesian authorities considered it a serious threat to the livestock industry in the Save Valley (Mavhunga and Spierenburg, 2007). In the 1940s, African cattle on the east bank of the Save River were affected and moved out to the north (Figure 3.3). Game elimination and bush clearing were also practised on the river's west bank (Ford, 1960). Study of tsetse habitats and spread in the 1940s and 1950s revealed the importance of both ungulate and human presence and mobility in the spread of tsetse. The area was situated along important labour migration routes towards South Africa (Murray, 1995; see also Chapter 6) and, '. . . the population on the Sabi River north of the [tsetse fly] advance has been and is a partial, if not complete barrier to the tsetse movement in this direction' (Ford, 1960, p. 6).

By 1962, tsetse flies had reached the confluence of the Chiredzi and Runde Rivers (Robertson *et al.*, 1972; Figure 3.6). Tsetse control measures now also started to include controls on human mobility (Mavhunga and Spierenburg, 2007),

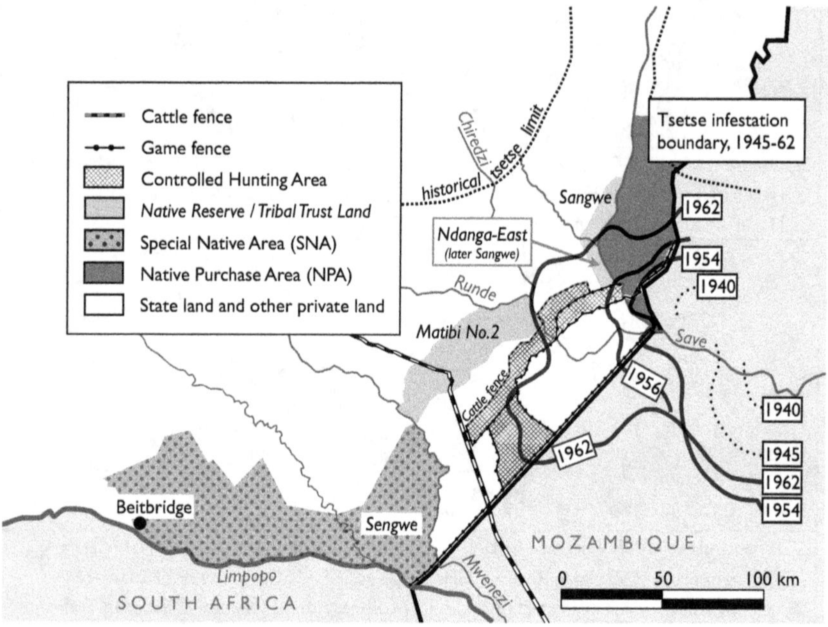

Figure 3.6 Tsetse and the fencing-off of people's resource access in the Southeast Lowveld, Southern Rhodesia, around 1964.

Sources: Ford (1960), Robertson *et al.* (1972) and Bossche and Vale (2000).

but game elimination and fences to control the movement of game and livestock mostly affected local people's resource access. An area to the north and south of the Runde River became surrounded by game fences and corridors. Within these fenced-in areas large mammals, the primary hosts of tsetse fly, were eliminated (Figure 3.6). On the northern boundary of the operational area, cattle fences were erected and a cattle-free corridor established. People and their cattle were moved north of the fence, which was located in the Sangwe Communal Land. This fence was to form the northern boundary of the newly created Gonarezhou National Park in 1975. The Chitsa people, who had been moved out to the north of this cattle-free corridor in the 1970s, reclaimed their land in the northern sector of the park in 2000 (Mombeshora and le Bel, 2009).

The tsetse flies were eventually eliminated from the Southeast Lowveld by 1970 through an intensive tri-national DDT spraying campaign that extended well into Mozambique. It prevented the further southward spread of the fly into the Mwenezi catchment and into South Africa's Kruger National Park.

Export markets and boundaries: Foot and Mouth Disease (FMD), 1931–2000

While tsetse infestation was an important driver of land tenure and protected area boundary formation in the early colonial period, the impact of FMD on boundary setting in future TFCAs was initially limited. Although an FMD outbreak on the Nuanetsi Ranch in Southern Rhodesia in 1931 had an enormous impact on Southern Africa's emerging overseas trade in frozen and chilled meat (Palmer, 1977; Phimister, 1978), FMD control measures did not result in the setting of new, permanent boundaries until decades later.

In 1931, FMD spread quickly from the massive and unfenced Nuanetsi Ranch, and resulted in Britain banning frozen beef imports from South Africa and Southern Rhodesia. It instigated an FMD control policy based on cattle movement restrictions (quarantines), the formation of cattle-free zones and vaccination (Condy, 1979). Such measures were generally temporary, seeking to contain FMD outbreaks, but affected both large-scale cattle ranchers and small-scale African cattle keepers.

> Even after quarantine is lifted stock are generally in such inferior condition that sales are undesirable. It is not only the districts in which the disease appears but all neighbouring areas that are quarantined and during the year Nuanetsi, Chibi, Ndanga, Bikita, Buhera, Melsetter, Chipinga, Umtali and part of Victoria were all at a standstill. These districts account for some 24% of all African owned cattle in the country.
>
> (Chief Native Commissioner, Annual report of 1953, Southern Rhodesia, 1946–1956)

FMD quarantine measures thus often undercut African livestock production. This was particularly the case in areas bordering (protected) wildlife areas, as the involvement of wildlife, buffalo (*Syncerus caffer*) in particular, in the spread of FMD became apparent (Condy, 1979). Outbreaks were generally confined to the Southeast Lowveld in Southern Rhodesia and the lowveld bordering the Kruger National Park in South Africa, both home to buffalo populations (Jori *et al.*, 2009; Foggin, 2010). The need to limit contact between cattle and buffalo in order to control FMD reinforced protected area boundaries.[17] Rather than conservationist concerns, government initiative to fence parks and game reserves, such as the Kruger National Park in the late 1950s (Box 3.1), was thus instigated by beef export considerations. The governments' serving the interests of the livestock industry is also apparent in Zimbabwe's buffalo destruction regulations (in 1979 and 1986). They followed cattle ranchers' calls for the eradication of buffalo from farming areas (Foggin, 2010, p. 77).

Global trade and FMD zoning: fences and livelihoods in Zimbabwe's Southeast Lowveld

Exporting beef to overseas markets requires control of FMD to conform to the regulations of the World Organization for Animal Health (OIE). These regulations are based on the concept of confining FMD to infected zones (red), with an FMD-vaccinated zone (green) acting as a buffer between infected zones and FMD-free zones, and restricting cattle and wildlife movements between these zones (Jori *et al.*, 2009, p. 918; Burroughs, 2010; Foggin 2010). Cattle and beef exports are only allowed from FMD-free zones, and cattle and game fences separate the disease management zones. Zimbabwe's independence in 1980 ended economic sanctions and opened up European meat markets to it. Such international markets as well as preferential trade agreements of the European Union resulted in a proliferation of veterinary fences, marking the boundaries between the different FMD zones, not only in Zimbabwe but in other Southern African countries as well (Ross, 2010).

Export-oriented cattle production had two major consequences for people living on the edge of protected wildlife areas where FMD is endemic. First, FMD zoning resulted in price differentials between cattle in the different zones, the price being higher in FMD-free zones (Jori *et al.*, 2009, p. 925; Chaminuka, 2012). Thus, cattle keepers in communal areas and small-scale commercial farming (SSCF) areas in Zimbabwe's Southeast Lowveld are forced to sell at lower prices, which negatively affects those with cattle-oriented livelihoods (Murungweni, 2011). Second, FMD fencing has, in some cases, resulted in new boundaries. For instance, in Zimbabwe FMD fencing started in the 1980s, and resulted in EU-funded, buffalo-proof game fences stretching from Hwange National Park's eastern border to the Mid-Zambezi Valley and around Gonarezhou National Park (Foggin 2010, p. 77). The exercise resulted in new boundaries restricting resource access, as exemplified by the fencing of Gonarezhou National Park. Here, the fence makers excluded park area in the southwest, but included part of Ndowoyo Communal Area within the fence, and by following an existing track, also included the Naivasha Wilderness Area (Figure 3.7). This is an unoccupied part of Matibi

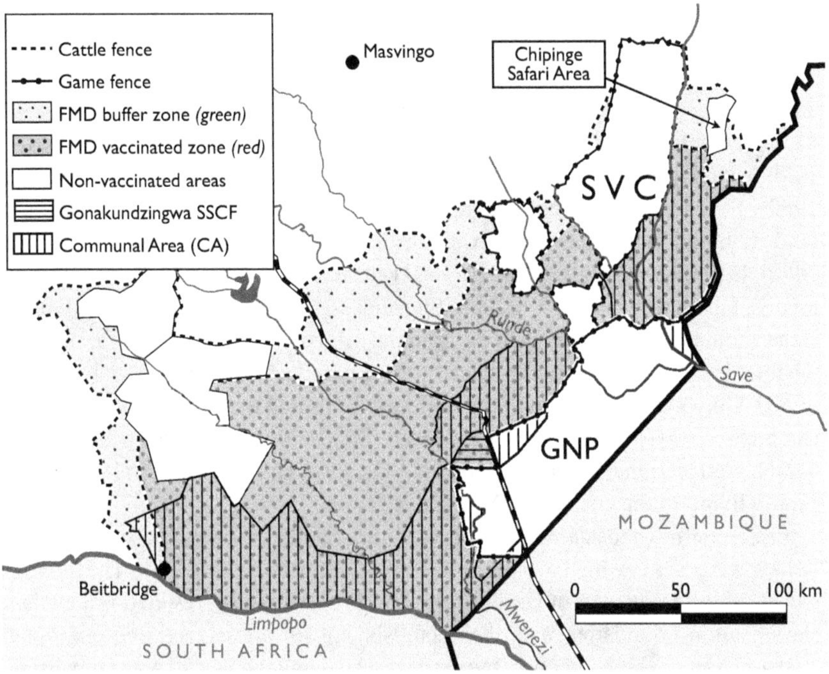

Figure 3.7 FMD in the Southeast Lowveld, Zimbabwe, late 1990s (SVC – Save Valley Conservancy).

Sources: Surveyor General (1988), Foggin (2010), authors' observation.

No. 2 Communal Area that remains under the control of the Chiredzi Rural District Council. In northern Zimbabwe, similar resource access restrictions resulted from fencing the boundaries of protected areas for disease control purposes (Box 3.4). As a consequence, veterinary fences have contributed to the contestation of land classification boundaries, resource access and benefit sharing.

In conclusion, the threat of diseases was not only an important driver of land use and tenure divisions in current TFCAs in the early twentieth century. Later, the markers of such divisions, fences, in turn drove new divisions. Boundary formation in relation to the threat posed by tsetse flies, and the spread of FMD exemplifies this. The alignment and realignment of fences has thus played a key role in defining the location and nature of the edge. Local peoples' views and needs have mostly been ignored or overridden in such fencing exercises (see also Ferguson and Hanks, 2010).

Eco-tourism and safari hunting in the post-World War II period

The game reserves and national parks that had sought to protect game and regulate hunting proved successful recreational destinations. Stimulated by international air travel, the post-World War II period saw an expansion of private and state-owned game reserves in Southern Africa (for example, Box 3.1 and Figure 3.2). While governments' central role in game protection (wildlife belonged to the state) and a strong political lobby of settler agriculture (fearing livestock disease) had hampered wildlife utilization initiatives on private lands (Child, B., 2009, p. 105), wildlife enterprises (hunting and game ranching) started to emerge during the 1950s. For instance, in South Africa, the state moved towards a privatization of wildlife by allowing hunting on well-fenced farms (Muir-Leresche and Nelson, 2000). In Southern Rhodesia and colonial Mozambique, the state got more actively involved in wildlife utilization. A hunting concession system in designated wildlife areas (coutadas) was instituted in colonial Mozambique. As these areas were inhabited, local people's small-scale hunting practices continued largely unsanctioned. Only a post-civil war surge in donor funds and the formation of the Great Limpopo Transfrontier Park in the early 2000s introduced the possibility of uninhabited national parks to Mozambique through physical displacement of people living in the coutadas (Dear and McCool, 2010).

In Southern Rhodesia, where sport hunting had been limited to private lands, CHAs were set up in the early 1960s (see Figures 3.3 to 3.5). The government's growing interest in wildlife production systems also resulted in new land classification boundaries with the establishment, in unassigned and communal land, of CHAs, which later became game reserves and then safari areas or national parks (Figure 3.5). Most of the Matetsi commercial farming area was also appropriated by the state in the early 1970s to form the Matetsi Safari Area and then leased to private safari operators (Cumming, 1989; Figure 3.4). Safari areas sought to preserve wildlife and natural habitats for the purpose of commercial exploitation (Parks and Wildlife Act, 1975). The safari industry grew rapidly, with safari operations on existing forest lands. New land use boundaries also resulted from safari operations being started on private and Tribal Trust Lands (as the former Native Reserves and many Special Native Areas became known) (Cumming, 1989). The largest safari areas were established in the northern part of the country (Box 3.4 and Figure 3.5). Hunting concessions in Tribal Trust Lands (later communal areas) were usually in sparsely or uninhabited areas, such as the former SNAs, bordering national parks or forest areas. But population growth elsewhere, tsetse eradication and resettlement programmes resulted in a rapid expansion of smallholder farming and domestic cattle herds in such areas in the post-war period. Smallholder cash crop production, notably cotton, constituted another driver of settlement in districts such as Gokwe, Hurungwe and Binga (Worby, 2000; Nyambara, 2001; Chimhowu, 2002), and the Mid-Zambezi Valley in the 1980s (Baudron *et al.*, 2011). As consequence, wildlife utilization in communal areas came increasingly under pressure. The number of hunting concessions on communal land was reduced in the 1980s, and as human presence depresses wildlife densities,[18] revenues from these areas are generally lower than from safari areas (Cumming, 1989). A Communal Areas Management Programme for Indigenous Resources (CAMPFIRE) was initiated in 1989 in order to involve people in these areas in wildlife management and to develop institutional mechanisms for benefit sharing and local development (Martin, 1986; Taylor and Martin, 1987). The lion's share of country-wide revenues from wildlife utilization did, however, continue to befall local government and private operators. Increasingly, cattle ranchers shifted to the lucrative wildlife business, as the formation of the Save Valley Conservancy exemplifies (Box 3.2 and Figure 3.3; du Toit 1992; Abel *et al.*, 2006). Meanwhile the benefits from CAMPFIRE programmes in communal areas have not been able to stem the increasing pressures on wildlife resources on the protected area edge. Nor have the revenues of these programmes been able to create sustainable development at the local level (Cumming and Lynam, 1997; Cumming, 2005).

Transfrontier conservation and boundary formation: participatory resettlement, zoning and wildlife corridor formation

While the advent of transfrontier conservation in Southern Africa has resulted in numerous TFCA-defining lines on maps, on the ground such boundaries often remain ill-defined and not marking clear discontinuities (Figure 1.2). TFCA formation has, thus far, not driven the formation of many new boundaries on the protected area edge. One exception is the resource use zoning within the Limpopo National Park that followed its establishment (see Figure 4.1). This zoning has resulted in a lengthy, and still ongoing, process of participatory resettlement of people living in the centre of the newly defined park (Milgroom, 2012).

A second exception involves the establishment of a wildlife corridor connecting the Gonarezhou and Kruger National Parks (Box 3.5, and Figures 1.2 and 3.3).

While such boundary formation has been driven by the highly political process that TFCA formation entails, implementation of these plans has seen various degrees of participation of people living on the edge. In Mozambique, this process is led by staff of the newly established Limpopo park. While in Zimbabwe, local government institutions, who have the jurisdiction of the land, have involved traditional leaders and local peoples in the planning and implementation process.

Conclusion

Geographical boundaries within Southern Africa's TFCA areas, whether in pre- or post-colonial times, have tended to be short lived in response to the ever-changing social and ecological dynamics of these areas. This chapter has shown that in order to understand boundary formation in current TFCAs, our focus needs to go beyond the two defining sets of boundaries of TFCAs – national and protected area borders. Capitalist expansion, disease and declining wildlife numbers were not the only important drivers of these two sets of boundaries in the nineteenth and early twentieth centuries. These two sets of boundaries have also been profoundly influenced by land tenure boundaries and disease infestation boundaries.

Box 3.5 Participatory Protected Area formation: the Sengwe Corridor in the GLTFCA, Zimbabwe

When Zimbabwe signed the treaty to establish the Great Limpopo Transfrontier Park (GLTFP) in late 2000, it also committed itself to establish a wildlife corridor linking the Kruger and Gonarezhou National Parks. This corridor could only be established through Sengwe Communal Land. Initially, corridors as wide as the northern boundary of Kruger National Park (KNP) (approx. 25 km) were shown on maps of the GLTFP (see also Figure 1.2). Such a corridor was unrealistic, as it included heavily settled areas within the Sengwe Communal Land, and would have involved relocating 5,600–15,500 people. Initially, it was assumed that the area would be incorporated into the Parks and Wildlife estate. As this idea found no favour locally, determining the extent and boundaries of the Gonarezhou–Kruger corridor became a consultative process, involving central government, the Chiredzi and Beitbridge Rural District Councils, traditional leaders and villagers. A series of meetings and workshops were held. Within Sengwe Communal Land, a TFCA committee was formed.

At the request of the Chiredzi Rural District Council (RDC), under whose jurisdiction Sengwe falls, CESVI developed a concept paper for the corridor, after consultations with the RDCs (Chiredzi and Beitbridge), the TFCA committee and local communities (CESVI, 2002). The concept paper recommended a corridor of between 4.8 and 8.4 km wide along the border with Mozambique between the Gonarezhou National Park (GNP) and KNP. It further recommended the establishment of a wildlife area, the Limpopo Strip, of 20–30 km along the Limpopo River and 0.6–2.3 km wide covering an area of about 25 km^2. Traditional leaders assisted in demarcating the boundary of the corridor and strip. With the southern bank of the Limpopo River lying within the Makuleke Contractual Park (i.e. the northern sector of KNP), the strip along the northern bank was intended to provide for a wildlife area covering both banks of the Limpopo River. This option would provide for the development of lodges along the river in a very attractive tourist area.

The two RDCs involved subsequently convened a series of workshops in the area. It became clear that government plans to incorporate the corridor into the GNP were not acceptable to the Sengwe people who feared loss of agricultural and grazing land. Early in 2003, a seminar on the collaborative management of the proposed Sengwe-Tshipise Corridor and Limpopo Strip examined management and legal options for establishing the corridor (Daconto 2003a). The RDCs then decided to have the proposed corridor declared a wilderness area within the communal land under the Regional, Town and Country Planning Act (Chapter 29:2) and the Traditional

Leaders Act (Chapter 29:17). Under these Acts, jurisdiction for the corridor and strip would remain with the Chiredzi and Beitbridge RDCs. A comprehensive participatory rural appraisal study of the ecological and social impacts of the corridor was undertaken and its results reviewed at a workshop in July 2003; and thereafter at report-back village workshops (Cunliffe 2003; Daconto 2003b). The formal planning process was then set in motion but the Sengwe-Tshipise Wilderness Area was only finally gazetted in December 2009.

While protected area formation was often born out of conservationist concerns, capitalist interests and disease considerations were never far away. Similar to current TFCA creation, the location and boundaries of the established game reserves and national parks were seldom the outcome of deliberate conservation planning, but were foremost political-administrative constructions. First, they were usually established in sparsely populated, disease-ridden marginal lands that white settlers and their governments had left 'unassigned' as they were deemed unfit for agricultural development. Protected areas were the 'leftovers' of a process of land alienation that resulted in a racial division of land with different tenure regimes. Land tenure boundaries constituted a defining set of boundaries in the settler economies of Southern Africa and thus of the future edge areas.

Second, the establishment of protected areas directly served the political and economic interests of the white settlers, often at the expense of Africans living in these areas. While the establishment of the first national parks such as Kruger were part and parcel of nation-building projects of young states, hunting and tourism in game reserves became important forex-earning economic sectors in the expanding capitalist economies of Southern Africa. Where the interests of its political supporters were threatened, the settler governments would come to their rescue. Thus we can understand the state support for struggling white livestock producers on the agriculturally marginal lands bordering the state-established 'disease reservoirs', the protected wildlife areas and its central role in livestock disease management and fencing.

Protected area formation thus did not usually result in large-scale resettlement of Africans, as these agriculturally marginal areas were generally sparsely populated (Child, G., 2009). Yet, the number of people living on the edge of these protected areas, and within current TFCAs, grew rapidly. Natural population growth and resettlement from more populated areas and settlers' private lands increased pressure on natural resources in these areas. While Chapter 4 takes up this important theme of increased population growth and its drivers in more detail, this chapter has shown that such growth and shifting disease boundaries reduced both the conservation value of the protected area edge, local peoples' access and their chances to benefit from wildlife resources.

Notes

1 This section of the boundary was reconfirmed by an exchange of ratifications in 1882 by Portugal and the United Kingdom, which had annexed the Afrikaner Republic in 1877. Self-government of the Afrikaner Republic was restored in 1881 when it became known as the Transvaal state (IBS, 1973).

2 The initial boundary was further adjusted and refined by arbitration and negotiation over an extended period with the final agreement being reached in 1940.

3 The coutadas that are now part of the Great Limpopo TFCA were later redefined as Zinave, Banhine (both in 1972) and Limpopo National Parks (in 2001) (see Figure 1.2). People living within them continued to reside there, and eco-tourism remained underdeveloped. Hence, the idea of national parks as protected areas devoid of people did not apply here (cf. Carruthers, 2009, p. 36). Only in the 2000s have there been attempts to relocate people living in these areas, or restrict their use of natural resources other than wildlife (Dear, 2008; Milgroom and Spierenburg, 2008; Dear and McCool, 2010).

4 The first two Native Reserves, the Gwaai and Shangani, are an exception. They were defined before the office of Native Commissioner had been created, and before the Matabele and Shona risings of 1896. Although claimed as good grazing land, the Ndebele considered the waterless lands as 'cemeteries, not homes'. The Shangani Reserve was also within the tsetse-infested area (Palmer, 1977, pp. 30–34).

5 Land committees included the Native Affairs Committee of 1910, the 1914–1915 Native Reserves Commission and the 1924–1925 (Morris Carter) Land Commission. The main recommendations of the latter were embodied in the Land Apportionment Act of 1930, which legalized racial segregation in land holding (Jennings and Huggins, 1935).

6 Not surprisingly, lands in NPAs were not taken up quickly, especially not in the remote, future conservation 'edge' areas. For instance, in Chibi District in the Southeast Lowveld, only four farms were sold by 1936, the same number as in the Lomagundi District in the north. In Matabeleland figures were even lower (Palmer, 1977, pp. 214–215).

7 What the official status of the area that now forms Gonarezhou National Park was from 1930 to the 1960s remains unclear.

8 Because of the long dry season and poor grass quality during the dry season the land was not in fact suitable for cattle production without protein supplements through the dry season – this problem was only recognized in the 1960s.

9 The rinderpest-induced receding of tsetse infestation has been challenged by Summers (1967, cited in Child and Riney, 1987) and Campbell and Child (1971), who have argued the tsetse boundary was already receding at the time of the pandemic.

10 It contributed to the near extinction of wildlife species like the black rhinoceros (*Diceros bicornis*) (Child and Riney, 1987, p. 12). Tsetse control hunting also contributed to the reduced growth of the country's elephant population in the 1960s and 1970s (Cumming, 1981).

11 Extensive game elimination followed an outbreak of trypanosomiasis on a European farm in the Gwaai River area in 1918. In 1923, the focus of intervention shifted to the Lomagundi District where both settler and African farmers were affected by the expansion of the tsetse-infested area (Jack, 1933, p. 369). In Chegutu and Kadoma, new settlers started to lose their cattle to disease in the late 1920s (Jack, 1933; Lovemore, 1994).

12 Strictly speaking, Wankie was not the country's first game reserve. The government proclaimed 224,000 acres in the Matopos, south of Bulawayo 'for the purpose of a National Park and Game Reserve' in 1926. Yet, the park had no staff (controlling hunting) and remained inhabited until the 1960s (Ranger, 1999, pp. 62, 136).

13 The London Convention defined a national park as an area 'set aside for the propagation, protection, and preservation of objects of aesthetic, biological, pre-historical, archaeological or other scientific interest' (Ranger, 1999, p. 137).

14 Centralization reorganized land in Native Reserves in consolidated blocks of grazing and arable lands, and villages along lines dividing arable and grazing lands (McGregor, 1995).

15 The initial forced displacements from European land, and resulting from the construction of the Kariba dam in the late 1950s, were followed by 'voluntary' migrations from different, densely populated parts of the country, particularly the south. 'The arrival of these immigrants coincided with the introduction of cotton production in Gokwe by state officials' (Nyambara, 2001, p. 773).

16 Urungwe Native District (1956) 'Outline scheme for the resettlement of natives in a tsetse fly area, 1956 onwards' document written by Land Development Officer R.R. Gladdon. See: www.sacema.com/uploads/tsetse/TG1/tsetse-general-92.pdf; see also: R. Goodier (1957) 'A preliminary report on the Rwenge Valley reclamation scheme, October 1957', www.sacema.com/uploads/tsetse/TG1/tsetse-general-100.pdf.

17 Cattle owners tried to limit the contact between cattle and buffalo not merely for this reason. Corridor disease (Theileriosis) is also transmitted from buffalo to cattle, through ticks, but unlike FMD, is generally fatal for cattle.

18 There is a seasonal component to these lower wildlife densities. Hunting takes place in the dry season when there are no crops in the field and limited forage. Animals are drawn to fields in the rainy season, when they cause damage to crops.

References

Abel, N., D.H.M. Cumming and J.M. Anderies (2006) 'Collapse and reorganization in social-ecological systems: questions, some ideas, and policy implications', *Ecology and Society*, vol. 11, no. 1. Available at: www.ecologyandsociety.org/vol11/iss1/art17/

Andersson, J.A. (1999) 'The politics of land scarcity: land disputes in Save Communal Area, Zimbabwe', *Journal of Southern African Studies*, vol. 25, no. 4, pp. 553–578

Bannerman, J. (1978) 'Towards a history of the Hlengwe people of the south of Rhodesia', *NADA*, vol. 11, no. 5, pp. 483–496

Baudron, F., M. Corbeels, J.A. Andersson, M. Sibanda, and K.E. Giller (2011) 'Delineating the drivers of waning wildlife habitat: the predominance of cotton farming on the fringe of protected areas in the Mid-Zambezi Valley, Zimbabwe', *Biological Conservation*, vol. 144, pp. 1481–1493

Bolding, A., J. Mutimba, P. van der Zaag and J.A. Andersson, (2003) 'Introduction', in A. Bolding, J. Mutimba and P. van der Zaag (eds), *Interventions in Smallholder Agriculture: Implications for Extension in Zimbabwe*, University of Zimbabwe Publishers, Harare, pp. 1–32

Bossche, P. van der and G. Vale (2000) *Tsetse and Trypanosomosis in Southern Africa*, Regional Tsetse and Trypanosomosis Control Programme for Southern Africa, Harare

Brown, K. (2005) 'Tropical medicine and animal disease: Onderstepoort and the development of veterinary science in South Africa, 1908–1950', *Journal of Southern African Studies*, vol. 31, no. 3, pp. 513–529

Bryden, H.A. (1903) 'The decline and fall of the elephant in Southern Africa', *Fortnightly Rev.*, vol. 79, 108, pp. 100–101

Burroughs, R. (2010) 'Current policies for epizootic disease containment in the Southern Africa sub-region', in K. Ferguson and J. Hanks (eds), *Fencing Impacts: A Review of the Environmental, Social and Economic Impacts of Game and Veterinary Fencing in Africa with Particular Reference to the Great Limpopo and Kavango-Zambezi Transfrontier Conservation Areas*, Mammal Research Institute, University of Pretoria, pp. 56–61. Available at: www.wcs-ahead.org/gltfca_grants/grants.html

Campbell, A. and G.F.T. Child (1971) 'The impact of man on the environment of Botswana', *Botswana Notes and Records*, vol. 3, pp. 91–110

Campbell, B., A. Mandondo, N. Nemarundwe, B. Sithole, W. De Jong, M. Luckert and F. Matose (2001) 'Challenges to proponents of common property resource systems: despairing voices from the social forests of Zimbabwe', *World Development*, vol. 29, no. 4, pp. 589–600

Carruthers, J. (1989) 'Creating a National Park, 1910 to 1926', *Journal of Southern African Studies*, vol. 15, no. 2, pp. 188–216

Carruthers, J. (1994) 'Dissecting the myth: Paul Kruger and the Kruger National Park', *Journal of Southern African Studies*, vol. 20, pp. 263–283

Carruthers, J. (1996) 'Defending Kruger's honour? A reply to Professor Hennie Grobler', *Journal of Southern African Studies*, vol. 22, no. 3, pp. 473–480

Carruthers, J. (2007) 'Influences on wildlife management and conservation biology in South Africa c.1900 to c.1940', *South African Historical Journal*, no. 58, pp. 65–90

Carruthers, J. (2009) 'National Parks in South Africa', in H. Suich and B. Child (eds), *Evolution and Innovation in Wildlife Conservation: Parks and Games Ranches to Transfrontier Conservation Areas*, Earthscan, London, pp. 35–49

CESVI (2002) 'Sengwe Corridor concept paper', Technical Report, Sustainable Development and Natural Resource Management in Southern Zimbabwe, Project CESVI/MAE/AID 5063, CESVI Zimbabwe Office, Harare

Chaminuka, P. (2012) 'Evaluating land use options at the wildlife–livestock interface: an integrated spatial land use analysis', Ph.D. thesis, Wageningen University, Wageningen

Child, B. (2009) 'Private Conservation in Southern Africa: practice and emerging principles', in H. Suich and B. Child (eds), *Evolution and Innovation in Wildlife Conservation: Parks and Games Ranches to Transfrontier Conservation Areas*, Earthscan, London, pp. 103–112

Child, G. (2009) 'The emergence of modern conservation practice in Zimbabwe', in H. Suich and B. Child (eds), *Evolution and Innovation in Wildlife Conservation: Parks and Games Ranches to Transfrontier Conservation Areas*, Earthscan, London, pp. 67–84

Child, G. and L. Chitsike (2000) '"Ownership" of wildlife', in H.T.T. Prins, J.G. Grootenhuis and T.T. Dolan (eds), *Wildlife Conservation by Sustainable Land Use*, Kluwer Academic Publishers, Dordrecht, pp. 247–266

Child, G.F.T. (1995) *Wildlife and People: The Zimbabwean Success*, WISDOM Foundation, Harare and New York

Child, G.F.T. and T. Riney (1987) 'Tsetse control hunting in Zimbabwe, 1919–1958', *Zambezia*, vol. 14, no. 1, pp. 11–71

Chimhowu, A. (2002) 'Extending the grain basket to the margins: spontaneous land resettlement and changing livelihoods in the Hurungwe district, Zimbabwe', *Journal of Southern African Studies*, vol. 28, no. 3, pp. 551–573

Chorley, J.K. (1938) 'A short historical account of tsetse in Southern Rhodesia', *Proceedings of the Rhodesian Scientific Association*, vol. 36, pp. 1–11

Condy, J.B. (1979) 'A history of foot and mouth disease in Rhodesia', *Rhodesian Veterinary Journal*, vol. 10, pp. 1–10

Cumming, D.H.M. (1981) 'The management of elephant and other large mammals in Zimbabwe', in P.A. Jewell, S. Holt and D. Hart (eds), *Problems in Management of Locally Abundant Wild Mammals*, Academic Press, New York, pp. 91–118

Cumming, D.H.M. (1989) 'Commercial and safari hunting in Zimbabwe', in R.J. Hudson, K.R. Drew and L.M. Baskin (eds), *Wildlife Production Systems: Economic Utilisation of Wild Ungulates,* Cambridge University Press, Cambridge, pp. 147–169

Cumming, D.H.M. (2005) 'Wildlife, Livestock and Food Security in the South-East Lowveld of Zimbabwe', in S.A. Osofsky, S. Cleaveland, W.B. Karesh, M.D. Kock, P.J. Nyhus, L. Starr and A. Yang (eds), *Conservation and Development Interventions at the Wildlife/Livestock Interface: Implications for Wildlife, Livestock and Human Health*, UICN, Gland and Cambridge, pp. 41–45

Cumming, D.H.M. and T.J.P. Lynam (1997) *Land Use Changes, Wildlife Conservation and Utilisation, and the Sustainability of Agro-Ecosystems in the Zambezi Valley: Final Technical Report*, European Union Contract B7-5040/93/06, WWF Programme Office, Harare

Cumming, G.S., D.H.M. Cumming and C. L. Redman (2006) 'Scale mismatches in social-ecological systems: causes, consequences, and solutions', *Ecology and Society*, vol. 11, no. 1, p. 14. Available at: www.ecologyandsociety.org/vol11/iss1/art14/

Cunliffe, R.N. (2003) *PRA Questionnaire Survey Study in Sengwe-Tshipise Communal Lands, Final Report, Part 1: Corridor Analysis*, CESVI Zimbabwe Office, Harare

Daconto, G. (2003a) *Proceedings of the Seminar on the Development of Collaborative Management for the Sengwe Corridor*, CESVI Zimbabwe Office, Harare

Daconto, G. (Compiler) (2003b) 'Report of the feedback workshop on the PRA study in the Sengwe-Tshipise Communal Lands', CESVI Zimbabwe Office, Harare

Davison, T. (1967) *Wankie: The Story of a Great Game Reserve*, Books of Africa, Cape Town

Dear, C.E. (2008) 'Causes and consequences of displacement decision-making in Banhine National Park, Mozambique', Ph.D. thesis, University of Montana, Missouala

Dear, C. and S. McCool (2010) 'Causes and consequences of displacement decision-making in Banhine National Park, Mozambique', *Conservation and Society*, vol. 8, no. 2, pp. 103–111

Delius, P. (1984) *The Land Belongs to Us*, University of California Press, Berkeley

DNPWLM (Department of National Parks and Wild Life Management) (1963–1976), Annual reports, Government Printers, Salisbury, Rhodesia

DWC (Department of Wildlife Conservation) (1962) *Annual Report of 1962*, Government Printers, Salisbury, Rhodesia

Dzingirai, V. (2003) '"CAMPFIRE is not for Ndebele migrants": the impact of excluding outsiders from CAMPFIRE in the Zambezi Valley, Zimbabwe', *Journal of Southern African Studies*, vol. 29, no. 2, pp. 445–459

Ferguson, K. and J. Hanks (2010) *Fencing Impacts: A Review of the Environmental, Social and Economic Impacts of Game and Veterinary Fencing in Africa with Particular Reference to the Great Limpopo and Kavango-Zambezi Transfrontier Conservation Areas*, Mammal Research Institute, University of Pretoria. Available at: www.wcs-ahead.org/gltfca_grants/grants.html

Floyd, B. (1962) 'Land apportionment in Southern Rhodesia', *Geographical Review*, vol. 52, no. 4, pp. 566–582

Foggin, C. (2010) 'Zimbabwe's livestock disease control policy in relation to fencing', in K. Ferguson and J. Hanks (eds), *Fencing Impacts: A Review of the Environmental, Social and Economic Impacts of Game and Veterinary Fencing in Africa with Particular Reference to the Great Limpopo and Kavango-Zambezi Transfrontier Conservation Areas*, Mammal Research Institute, University of Pretoria, pp. 75–82. Available at: www.wcs-ahead.org/gltfca_grants/grants.html

Ford, J. (1960) 'The Advance of *Glossina morsitans* and *Glossina pallidipes* into the Sabi and Lundi River Basins, Southern Rhodesia', 8th ISCTR meeting. Available at: www.sacema.com/uploads/tsetse/TG1/tsetse-general-66.pdf

GOZ (Government of Zimbabwe) (2007) *Zimbabwe Land Classification Map 1998*, Department of the Surveyor General, Harare

Grobler, H. (1996) 'Dissecting the Kruger myth with blunt instruments: a rebuttal of Jane Carruthers's view', *Journal of Southern African Studies*, vol. 22, no. 3, pp. 455–472

IBS (1964) *International Boundary Study No. 30: Zambia–Zimbabwe (Southern Rhodesia) Boundary*, The Geographer, Office of the Geographer, Bureau of Intelligence and Research, Department of State, USA

IBS (1971a) *International Boundary Study No. 117: South Africa–Zimbabwe (Southern Rhodesia) Boundary*, The Geographer, Office of the Geographer, Bureau of Intelligence and Research, Department of State, USA

IBS (1971b) *International Boundary Study No. 118: Mozambique–Zimbabwe (Southern Rhodesia) Boundary*, The Geographer, Office of the Geographer, Bureau of Intelligence and Research, Department of State, USA

IBS (1973) *International Boundary Study No. 133: Mozambique–South Africa Boundary*, The Geographer, Office of the Geographer, Bureau of Intelligence and Research, Department of State, USA

IBS (1974) *International Boundary Study No. 14: Botswana–Zimbabwe (Southern Rhodesia) Boundary*, The Geographer, Office of the Geographer, Bureau of Intelligence and Research, Department of State, USA

IEA (2010) Database Project, International Environmental Agreements, University of Oregon. Available online: http://iea.uoregon.edu/pages/view_treaty.php?t=1900-PreservationWildAnimalsBirdsFishAfrica.EN.txt&par=view_treaty_html (accessed 25 May 2012)

Jack, R. (1930) 'Discussions concerning game preservation, trypanosomiasis and tsetse fly', *Rhodesia Agricultural Journal*, vol. 27, pp. 827–840

Jack, R. (1933) 'The tsetse fly problem in Southern Rhodesia', *Rhodesia Agricultural Journal*, 30, pp. 365–384

Jennings, A. and G. Huggins (1935) 'Land apportionment in Southern Rhodesia', *Journal of the Royal African Society*, vol. 34, no. 136, pp. 296–312

Jori, F., W. Vosloo, B. Du Plessis, R. Bengis, D. Brahmbhatt, B. Gummow and G.R. Thomson (2009) 'A qualitative risk assessment of factors contributing to foot and mouth disease outbreaks in cattle along the western boundary of the Kruger National Park', *OIE Revue Scientifique et Technique*, vol. 28, no. 3, pp. 917–913

Lindsey, P., R. du Toit, A. Pole and S. Romañach (2009) 'Save Valley Conservancy: a large-scale African experiment in cooperative wildlife management', in H. Suich and B. Child (eds), *Evolution and Innovation in Wildlife Conservation: Parks and Games Ranches to Transfrontier Conservation Areas*, Earthscan, London

Livingstone, D. (1857) *Missionary Travels and Researchers in South Africa*, Gutenberg ebook. Available at: www.gutenberg.org/ebooks/1039

Lovemore, D.F. (1994) 'The influence of tsetse distributions on settlement and land utilisation in Zimbabwe, with reference to the effect to the EC funded regional tsetse and trypanosomiasis control programme (RTTCP) of Malawi, Mozambique, Zambia and Zimbabwe', *A Systematic Approach to Tsetse and Trypanosomiasis Control*, FAO, Rome.

McGregor, J. (1995) 'Conservation, control and ecological change: the politics and ecology of colonial conservation in Shurugwi, Zimbabwe', *Environment and History*, vol. 1, no. 3, pp. 257–279

Mackenzie, J.M. (1988) *The Empire of Nature: Hunting, Conservation and British Imperialism*, Manchester University Press, Manchester

Mavhunga, C. and M. Spierenburg (2007) 'A finger on the pulse of the fly: hidden voices of colonial anti-tsetse science on the Rhodesian and Mozambican borderlands, 1945–1956', *South African Historical Journal*, vol. 58, pp. 117–141

Martin, R.D. (1986) 'The Communal Areas Management Programme for Indigenous Resources', Department of National Parks and Wild Life Management, Harare

Meredith, M. (2008) *Diamonds, Gold and War: The Making of South Africa*, Jonathan Ball Publishers, Johannesburg

Milgroom, J. (2012) 'The elephants of democracy: an unfolding process of resettlement and accessing natural resources in the Limpopo National Park', Ph.D. thesis, Wageningen University, Wageningen

Milgroom, J. and M. Spierenburg (2008) 'Induced volition: resettlement from the Limpopo National Park, Mozambique', *Journal of Contemporary African Studies*, vol. 26, no. 4, pp. 435–448

Milton, S. (1997) 'The Transvaal Beef frontier: environment, markets and the ideology of development, 1902–1942', in T. Griffiths (ed.), *Ecology and Empire: Environmental History of Settler Societies*, Keele University Press, Edinburgh, pp. 199–211

Mombeshora, S. and S. le Bel (2009) 'Parks-people conflicts: the case of Gonarezhou National Park and the Chitsa community in south-east Zimbabwe', *Biodiversity and Conservation*, vol. 18, pp. 2601–2623

Muir-Leresche and R.H. Nelson (2000) *Private Property Rights to Wildlife: The Southern African Experiment*, CEI discussion paper. Available online: www.cei.org

Murphree, M.W. and D.H.M. Cumming (1993) 'Savanna land use: policy and practice in Zimbabwe', in M.D. Young and O.T. Solbrig (eds), UNESCO/Parthenon Publishing Group, Paris, pp. 139–178

Murungweni, C. (2011) 'Vulnerability and resilience of competing land-based livelihoods in south eastern Zimbabwe', Ph.D. thesis, Wageningen University, Wageningen

Mutowo, M. (2001) 'Animal diseases and human populations in colonial Zimbabwe: the rinderpest epidemic of 1896–1898', *Zambezia*, vol. 28, no. 1, pp. 1–22

Mutwira, R. (1989) 'Southern Rhodesian Wildlife Policy, 1890–1953: a question of condoning game slaughter?' *Journal of Southern African Studies*, vol. 15, no. 2, pp. 250–262

Ndlovu-Gatsheni, S.J. (2009) 'Mapping cultural and colonial encounters, 1880s–1930s', in B. Raftopoulos and A.S. Mlambo (eds), *Becoming Zimbabwe: A History from the Pre-Colonial Period to 2008*, Weaver Press, Harare

Nyambara, P. (2001) 'Immigrants, "traditional" leaders and the Rhodesian state: the power of "Communal" land tenure and the politics of land acquisition in Gokwe, Zimbabwe, 1963–1979', *Journal of Southern African Studies*, vol. 27, no. 4, pp. 771–791

Pakenham, T. (1991) *The Scramble for Africa*, Abacus, London

Palmer, R. (1977) *Land and Racial Domination in Rhodesia*, Heinemann, London

Phimister, I. (1978) 'Meat and monopolies: beef cattle in Southern Rhodesia, 1890–1938', *Journal of African History*, vol. 19, no. 3, pp. 391–414

Phimister, I. (1988) *An Economic and Social History of Zimbabwe, 1890–1948*, Longman, London

Phimister, I. (1993) 'Rethinking the reserves: Southern Rhodesia's Land Husbandry Act reviewed', *Journal of Southern African Studies*, vol. 19, no. 2, pp. 225–239

Pienaar, U. de V. (1978) 'The Kruger Park Saga: 1898–1981', unpublished manuscript

Pollard, S., C. Shackleton and J. Carruthers (2003) 'Beyond the fence: people and the lowveld landscape', in J.T. Du Toit, K.H. Rogers, and H.C. Biggs (eds), *The Kruger Experience: Ecology and Management of Savanna Heterogeneity*, Island Press, Washington DC, pp. 422–446

Ranger, T. (1999) *Voices from the Rocks: Nature, Culture and History in the Matopos Hills of Zimbabwe*, James Currey, Oxford

Robertson, A.G., E.B. Kluge, D.A. Kritzinger and A.E. De Sousa (1972) 'The use of residual insecticides in the reclamation of the Rhodesia-Mozambique border region the Sabi/Save and Limpopo Rivers from *Glossina morsitans* Westwood', *Proceedings and Transactions of the Rhodesia Scientific Association*, vol. 55, no. 1, pp. 34–62

Robins, S. and K. van der Waal (2008) '"Model tribes" and iconic conservationists? The Makuleke restitution case in Kruger National Park', *Development and Change*, vol. 39, no. 1, pp. 53–72

Roe, E. (1995) 'More than the politics of decentralization: local government reform, district development and public administration in Zimbabwe', *World Development*, vol. 23, no. 5, pp. 833–843

Ross, K. (2010) 'The advocacy role of NGOs in promoting dialogue as a means of resolving conflicts resulting from veterinary cordon fencing in Botswana', in K. Ferguson and J. Hanks (eds), *Fencing Impacts: A Review of the Environmental, Social and Economic Impacts of Game And Veterinary Fencing in Africa with Particular Reference to the Great Limpopo and Kavango-Zambezi Transfrontier Conservation Areas*, Mammal Research Institute, University of Pretoria, pp. 40–43

Ross, R. (1997) *A Concise History of South Africa*, Cambridge University Press, Cambridge

SAHO (undated) 'Colonial history of Nelspruit', South African History Online. Available online: www.sahistory.org.za/print/nelspruit/colonial-history-nelspruit (accessed 23 November 2011)

Saunders, C. (2006) *Gonarezhou: A Place for Elephants*, Lowveld Lodge Enterprises, Mutare, Zimbabwe

Selous, F.C. (1909) 'Big game in South Africa and its relation to the tsetse fly', *Journal of the Royal African Society*, vol. 8, no. 30, pp. 113–129

Southern Rhodesia (1935), *Land Apportionment Map* (1st edition, scale 1: 1,000,000), Surveyor General, Salisbury, Rhodesia

Southern Rhodesia (1939) *Report of Commission to Enquire into the Preservation, etc., of the Natural Resources of the Colony*, Government Printer, Salisbury, Rhodesia

Southern Rhodesia (1943), *Land Apportionment* (3rd edition, scale 1: 1,000,000), Survey Unit Reserve Section, Surveyor General, Salisbury, Rhodesia

Southern Rhodesia (1946–1956) *Annual Report of the Director of Native Agriculture*, Government Printer, Salisbury, Rhodesia

Southern Rhodesia (1953) *Annual report of the Director of Native Agriculture*, Government Printer, Salisbury, Rhodesia

Southern Rhodesia (1957) *Land Apportionment Map*, Federal Government Printer, Salisbury

Southern Rhodesia (1968) *Land Apportionment* (revised to January 1965, scale: 1: 2,150,000), Surveyor General, Salisbury, Rhodesia

Surveyor General (1924) *Map of Southern Rhodesia, to Accompany Report of the Land Commission, 1925*, Government Printer, Salisbury, Rhodesia

Surveyor General (1988) *Veterinary Foot and Mouth Disease Control Fences*, Government Printer, Harare

Surveyor General (1998) *Zimbabwe Land Classification: Scale 1:1,000,000*, Government Printer, Harare

Taylor, R. and R. Martin (1987) 'Effects of veterinary fences on wildlife conservation in Zimbabwe', *Environmental Management*, vol. 11, no. 3, pp. 327–334

Toit, R.F. du (1992) 'Large-scale wildlife conservancies in Zimbabwe: opportunities for commercial conservation of endangered species', in W. van Hoven, H. Ebedes and A. Conroy (eds), *Wildlife Ranching: A Celebration of Diversity*, South African Game Organization, Pretoria, pp. 295–300

Wesseling, H.L. (1996) *Divide and Rule: The Partition of Africa, 1880–1914*, Praeger, New York

Wet, C. de (1989) 'Betterment planning in a rural village in Keiskammahoek, Ciskei', *Journal of Southern African Studies*, vol. 15, no. 2, pp. 326–334

Wolmer, W. (2003) *From Wilderness Vision to Farm Invasions: Conservation and Development in Zimbabwe's South-East Lowveld*, James Currey, London

Worby, E. (2000) '"Discipline without oppression": sequence, timing and marginality in Southern Rhodesia's post-war development regime', *Journal of African History*, vol. 41, no. 1, pp. 101–125

4 Population and livelihoods on the edge

Ken E. Giller, Frédéric Baudron, Steven Matema, Jessica Milgroom, Chrispen Murungweni, Chloé Guerbois and Wayne Twine

> The amount of land in Africa that's set aside for animals is tiny, maybe 3%[1] of Africa, and before Peace Parks was dwindling, and it could only go one way as population grew and grew. And yet, what is Africa? Africa is its animals . . . and that's the beauty of Africa – that's what makes it different from the rest of the world and to lose those animals would be catastrophic, so the wonderful thing about Peace Parks is trying to increase the amount of land for the animals, and by increasing the amount of land for the animals that will help human beings. The elephants are growing in numbers and . . . the last thing you would want to do is cull them, so what you ideally need to do is expand the land and space for the elephants.
>
> (Sir Richard Branson, Ambassador for Peace Parks' Elephant Corridor between Botswana and Zambia)[2]

Introduction

The above quote from Sir Richard Branson typifies popular ideas of wilderness and wildlife conservation in Africa that put animals before people. Other publicity surrounding peace parks suggests that local communities are likely beneficiaries from the developments of joining conservation areas across national boundaries (Katerere *et al.*, 2001; Dzingirai, 2004). New opportunities in terms of eco-tourism and job creation for local people are envisaged. This creates visions of a true 'win–win' situation where transfrontier conservation areas (TFCAs) promote both conservation and rural development. A 'honey-pot' effect has been described where national park creation leads to disproportionate population growth on the edge as a result of emerging opportunities associated with conservation areas (Wittemeyer *et al.*, 2008).

This chapter develops a different perspective. Against a backdrop of remarkable population growth over the past century in the region – for example the population of Zimbabwe grew from approximately half a million people in the early 1900s to around 13 million in 2002 (Cumming and Lynam, 1997) – we explore the factors that give rise to differences in population on the edge. We find that current population on the edge is the complex result of past and present (re)settlement policy, and rural development has not been spurred by conservation initiatives.

The analysis focuses on the drivers of changes in population at the edge, ranging from segregationist (apartheid) policies persistent in Southern Africa to spontaneous migration towards or away from the edge resulting from emerging opportunities. Such population movements not only lead to changes in pressure on natural resources on the edge, but also result in the coming together of ethnically diverse peoples with distinct socio-cultural orientations and livelihood pursuits. We explore the factors that have influenced population growth surrounding the conservation areas in the past, and current livelihood pursuits of people living on the edge. Through a number of examples from the edges of diverse conservation areas, we show how a huge range in livelihood pursuits and opportunities between locations has led to a wide variety of challenges for future developments of the conservation areas and of local livelihoods.

We start by introducing the agro-ecological similarities and heterogeneity of peoples, household composition and livelihood options at various scales – across countries, within regions, between villages and within local 'communities' (see Chapter 5 for a detailed discussion of 'community' and ethnic heterogeneity). We then describe the main activities that comprise the livelihoods of households living close to the edge, using specific examples that illustrate the particular opportunities that lead to people depending more, or depending less, on natural resources. Finally, we reflect on the diverse livelihoods of people living on the edge, which are shaped by ecological, socio-political, economic and institutional factors.

Agro-ecological similarities and local livelihoods

We focus our discussion on the local people who live within and on the edges of TCFAs in Southern Africa, namely the Great Limpopo Transfrontier Conservation Area (GLTFCA),[3] the Kavango-Zambezi (KAZA)[4] Park and the Lower Zambezi-Mana Pools Transfrontier Conservation Area[5] or Zambezi Heartland. Although some of these initiatives remain plans at present, local people find themselves living on the borders of protected areas within extensive TFCAs, and in the case of Limpopo National Park (LNP) within the boundaries of the newly designated park. The case studies presented in this chapter are selected to draw out important contrasts between countries, and locations within countries. They are chosen on the basis of availability of information from detailed studies and represent much of the variability in settlements found around the TFCAs.

Comparison of the agro-ecological characteristics of the different case study areas shows some remarkable similarities (Table 4.1). All of the areas are found at relatively low altitude (below 900 metres above sea level for the greater part, and often only 400–600 masl), with hot climates (mean annual temperature around 22–25°C) and poor and erratic rainfall (on average 400–600 mm) with a high risk of crop failure. In many of the areas farmers indicate they achieve a good harvest only once every four years. Soils are often sandy, stony and infertile on interfluves or heavy, naturally fertile black vertisols that are difficult to manage with hand tillage or animal traction. On the flood plains along river courses cropping fields are often concentrated on the more fertile alluvial soils. In most of the countries, the national land use classification potential indicates that the areas are suitable only for extensive livestock ranching. But despite these similarities – that situate these areas on the agro-ecological 'edge' of their respective countries in terms of environment – the current population densities of the areas vary enormously (Table 4.2) as do peoples' livelihood pursuits on the edge. Population density varies from only 2 to 10 people per km^2 along the edges in Mozambique to as much as 300 people per km^2 to the west of the Kruger National Park (KNP) in South Africa. A brief historical analysis can shed some light on the underlying reasons for these differences.

The Great Limpopo Transfrontier Conservation Area

Established in 2002, the Great Limpopo Transfrontier Conservation Area (GLTFCA) encompasses a huge area (*c.*100,000 km^2) straddling the borders of Zimbabwe, South Africa and Mozambique. In South Africa, the GLTFCA covers the Kruger National Park and several provincial and private nature reserves and game reserves on the western boundary (Chapter 1) and some communal areas are negotiating to have parts of their land included within the 'Kruger boundary fence'. In Mozambique and Zimbabwe, some 30,000 km^2 of land outside the national parks are included within the area of the GLTFCA. Thus literally tens of thousands of people find themselves living within proposed TFCAs.

Mozambique

The LNP was formed as one of the first steps in shaping the GLTFCA. One of the debates during the establishment of the LNP was contestation of the history of human settlement. These arguments claim that current settlement in the LNP is mostly post-civil war (after 1992), that population is growing in an area that is otherwise 'wild', and there is a need to conserve the land before it is 'ruined' by people. It was also argued that because the habitation of people is recent, current residents have no long-term rights to the land (Milgroom and Spierenberg, 2008), and no deep ancestral ties to specific locations (Witter, 2010).

At a broader scale, it has been established that humans have been living in the area since around 1000 AD. War, drought, floods and labour opportunities in South Africa have indeed moulded a history of mobility in the area that is now the LNP. Migrations between 1000 and 1500 AD led to the settlement of the Tsonga people along the southern Limpopo (Junod, 1962; Smith, 1973), who were then displaced by Nguni encroachments in the nineteenth century[6] (Harries, 1989). Witter (2010) argues that some of the residents of the Limpopo settled in their current locations in an attempt to evade Gaza Nguni domination (p. 77).

Natural resource dependency and population density in this area were already influenced by other economic activities in the nineteenth century. In the 1870s labour migration from the area that is now the LNP to South Africa was already common (Harries, 1994; Newitt, 1995). Initially, such migration was to diamond mines

Table 4.1 Major agro-ecological characteristics in the six study areas (data from Milgroom, Twine, Murungweni, Poshiwa, Baudron, Matema, Guerbois)

Variable	*Close to Limpopo National Park, Mozambique*	*Bushbuckridge, bordering Kruger National Park, South Africa*	*Southeast Lowveld, Zimbabwe, bordering Gonarezhou*	*Mid-Zambezi Valley, Zimbabwe*	*Tsholotsho, close to Hwange National Park, Zimbabwe*	*Hwange Rural District, Northern boundary of Hwange National Park, Zimbabwe*
Altitude (m above sea level)	109–215	800–450	200–500	400	800–1,000	900–1,100
Mean annual temperature (°C)	24 (max 32, min 16)	22	24 (max 30.9, min 16.7)	25		25
Total annual rainfall (mm)	400 (100–800)	From 550 mm in the east to 1,200 mm on the escarpment in the west	435 (excluding cyclone years (1999 and 2000), 645mm (including cyclone years)	650 (350 and 1,300)	590	606 (CV = 25%) (1928–2005)
Dominant soil types	Eutric fluvisols and mollic fluvisols along the rivers, haplic luvisols and arenosols, are the dominant soils along the edge of the park outside the river valleys	Predominantly shallow lithosols derived from potassic granites and grandiorite. Duplex soils and sodic clays in bottomlands. Vertic clays in a narrow gabbro dyke in the east	Eutric vertisols, eutric fluvisols, chromic luvisols, leptosols and ferralitic arenosols beginning with the most dominant	Sandy clayey loamy soils along major river and clayey interfluves, locally sodic	Kalahari sands (ferralitic arenosols), some aridic arenosols	Predominantly Kalahari sands, Batoka basalts (Sinamatella–Robins Areas) and Karoo sediments
Major vegetation types	*Colophospermum mopane* woodland, riparian forest dominated by Acacia spp.	Granite lowveld tall shrubland dominated by *Terminalia sericea*, *Sclerocarya birrea*, *Dichrostachys cinerea* and *Combretum* species. Legogote Sour Bushveld *Parinari curatellifolia* and *Bauhinia galpinii* on east slopes of the escarpment	*C. mopane*, *Combretum*. Acacia scrubland in low-lying areas	Deciduous dry savanna dominated by *C. mopane* trees	*Mopane*, *Combretum*. Acacia scrubland in low-lying areas	Mixed woodlands dominated by *Baikiaea plurijuga* and *T. sericea*. *C. mopane*, mostly on basaltic soils

Note: CV = coefficient of variation

Table 4.2 Major socio-economic indicators and farm characteristics in the study areas (data from Milgroom, Giva, Chaminuka, Twine, Murungweni, Chirozva, Poshiwa, Baudron, Matema, Guerbois)

Variable	*Close to Limpopo National Park, Mozambique*	*Bushbuckridge, bordering Kruger National Park, South Africa*	*Southeast Lowveld, Zimbabwe, bordering Gonarezhou*	*Mid-Zambezi Valley, Zimbabwe*	*Tsholotsho, close to Hwange National Park, Zimbabwe*	*Hwange Rural District, Northern boundary of Hwange National Park, Zimbabwe*
Population density (inhabitants per km^2)	From 2 (inside the LNP) to 7 along the Olifantes and Limpopo Rivers (4.8 average)	From 146 in the east to 303 in the west	From 9 in Sengwe across (Ward 14) up to 50 in areas close to Limpopo River, 34 in Chikombedzi (Ward 11)	17.3		From 10 ± 3 (Ward 8, 9, 12, 13) to 50 ± 7 (Ward 15, 16, 17) in 2010
Household (hh) size (No. per hh)	9	5.9 (excluding migrant members)	7.14 in communal and 11 in Gonakudzingwa SSC	6	6.18 (Ward 1, 3, 4, 7)	7.4 (data from Ward 15 in 2010)
Sex ratio (# men per 100 women)	75 (1997) for Gaza Province 90 (2010) for Massingir District	93	75	88 in 1992		90 in 2010 (Ward 15)
Ethnic groups	Shangaan	Shangaan, North Sotho	Shangaan, Venda and Hlengwe dominant (Karanga and Ndebele)	Doma, Korekore, Karanga	Kalanga, Ndebele (and few San)	Ndebele, Nambiya, Tonga, mainly
Population growth over last decade	22.7% increase since 1997	7.5% increase between 1992 and 2004		97% between 1992 and 2002		R_{pop} = 0.05 for the district between 2002 and 2010 (max 0.11 for Ward 16)
Farm size (ha)	2 to 20 ha per household. Very few commercial farms in the area	Large private game reserves in the east and north. No private ownership on communal lands	Gonakudzingwa SSC up to 750 ha Communal areas 3.7 ha (on average) ranging from 0.5 to 10 ha	3.4 ha	3.5–5.0 ha	1.2 ha per household on average (data from Ward 15 only)
Livestock ownership	74% have cattle, 64% have goats/sheep	28% own goats, 26% own cattle	47% own up to 20 cattle, 12% own over 20 cattle, donkeys	Cattle, sheep, donkeys	65% have cattle and donkeys, almost all have some goats or sheep	60% have cattle, 62% goats, 15% sheep and 6% donkeys (Ward 15 only)
Major crops	Maize, groundnut	Maize, groundnut	Sorghum, maize, groundnut, cowpea, watermelon and sweet sorghum, little cotton	Sorghum, maize, groundnut, cotton	Maize, sorghum, pearl millet and groundnut. Minor crops include cowpea, Bambara groundnut, sunflower and cotton. Melons and pumpkins are intercropped	Maize, sorghum, pearl millet, groundnut and Bambara groundnut

(Kimberly), collieries (Natal, Witbank) and sugar plantations (in both South Africa and Mozambique), which preceded migration to the goldmines on the Rand from the 1880s. The Witwatersrand Native Labour Association (WNLA), the recruitment arm of the South African mines, had a profound influence on the Mozambican 'edge' of what is now the LNP (Ranger, 1979). Although not the initiator of organized migration, from 1918, the WNLA established bases in Pafuri and Mapai, from which they recruited as far afield as Inhambane and Vilanculos. The WNLA developed roads and rest camps for the recruits[7] who walked through and camped in the area now designated as the LNP and KNP on their way to the mines. Thus an area of Mozambique that was regarded as uninhabited by the Portuguese was used by the WNLA as a 'labour reserve' for the mines.

The Portuguese later established agreements with the Union of South Africa to control and benefit from this outflow of labour by instituting policies that made labour an obligation, fees to travel to the mines, and a system of taxes and deferred payment to encourage the return of workers to Mozambique (Newitt, 1995). Because of its proximity to South Africa, the residents of the area of the LNP became oriented towards labour migration to the mines and urban centres and the majority of families had at least one member working in South Africa. This labour migration exerted a strong influence on the way people lived. Money earned in the mines allowed the rapid accumulation of cattle. The long-term absence of men resulted in a shortage of labour for cropping, which contributed to the shift from sorghum to maize. The labour shortage was exacerbated as colonial rule forced the remaining men to work for them on cotton plantations (Mandlate and Machele, 2009).

Another strong influence on the area was the drive for wildlife conservation. In response to a decline in wildlife numbers and South Africa's pressure to control hunting of valuable species in the 1920s, the Portuguese government set up a hunting concession known as Coutada 16 (established officially in 1960, but managed as a reserve since the 1930s) that later became the LNP (Mavhunga and Spierenberg, 2007). Coutada 16 (and today the LNP) shared boundaries with the Kruger National Park and was purposefully situated there to benefit from the overflow of game. Only white people had access to hunting permits, but many local residents worked as scouts (Witter, 2010). There was minimal control over local hunting for food.

Independence from Portugal in 1975 brought yet another wave of changes. In response to the Malawian government cutting off its state-organized labour supply in 1974, the mines reduced their recruitment of foreign labour to become less dependent on political decisions of countries hostile to the apartheid regime in South Africa (Davies and Head, 1995; Andersson, 2006). Recruitment of Mozambican workers for the mines declined. FRELIMO (the National Liberation Front of Mozambique) established state farms in each region and communal gardens in each village (Lunstrum, 2008). The FRELIMO policy of 'villagization' forced dispersed households to move into villages with organized neighbourhoods for the sake of economic and political control, but also to facilitate the provision of services, such as schools, hospitals and water pumps. The war with RENAMO (the National Resistance Movement) impacted the region of the LNP strongly in the mid- to late 1980s. About two-thirds of the population fled to South Africa. The remainder took refuge in protected villages, which were FRELIMO strongholds, although the area was controlled by RENAMO. Fields were burned, cattle were killed and normal life was not resumed until after the peace accords were signed in 1992. At this point the FRELIMO government made a significant effort to repatriate refugees by offering them transportation home and assistance with rehabilitation of their livelihoods. Post-civil war socialist policies concerning villagization were less rigid, and people were allowed to return to their original homesteads and no longer forced to stay in the communal villages, although many did.

In 2001, the LNP was declared to harmonize the status of the protected area that was to form part of the Great Limpopo Transfrontier Park (GLTFP) and the GLTFCA[8]. Given that Zimbabwe and South Africa already had national parks, the Mozambican government was urged to quickly establish one on their side of the border. Hunting and other resource harvesting for commercial use (such as charcoal) became illegal. Promises of jobs and economic development were made, but little materialized and it was unclear what was to become of the 27,000 people living within the LNP boundaries. The park was zoned to separate areas purely for wildlife, multi-use areas and zones for tourism of low, medium and high intensity (see Figure 4.1). The Shingwedzi River basin was considered ecologically valuable for wildlife and ideal for tourist development, which was deemed incompatible with activities such as cattle keeping and cropping. Therefore, it was decided in 2003 that the 7,000 residents living along the banks of the Shingwedzi River that runs through the centre of the park (Figure 4.1) would have to be resettled. The target locations are within the support zone of the LNP along the Limpopo River, or outside the park along the Rio Olifants (Milgroom and Spierenberg, 2008). By 2011 only one village of 18 families had been resettled, calling into question whether the overall resettlement plan will be realized (Milgroom, 2012).

Throughout the history of settlement, dependence on natural resources for livelihoods and the contribution from agriculture has varied, due to droughts, floods, insect plagues and livestock diseases (Liesegang, 1983; Newitt, 1995). Most people currently living inside the park had to move their homes as a result of the floods associated with Cyclone Eline in 2000. Agriculture is rain-fed and both the flood plains and upland fields are used in the rainy season. Dry season cropping exploits residual moisture in the lowland fields or small dips in the landscape that accumulate water. Fields are large and repeated sowing as well as spatial distribution of cropping areas is important for mitigating the risk of climate variability. Remittances from family members working in South Africa, though still common, are no longer as important as in the past. Today the combination of cropping and livestock provides the backbone for most families' sources of food, depending first on cropping, with sale of livestock as a fallback. Families have slowly rebuilt their herds since the war ended in 1992. Some villages have a cooperative where irrigation is possible on small plots, and there are a handful of private farmers who have equipment for irrigation, but the majority of agriculture is rain-fed.

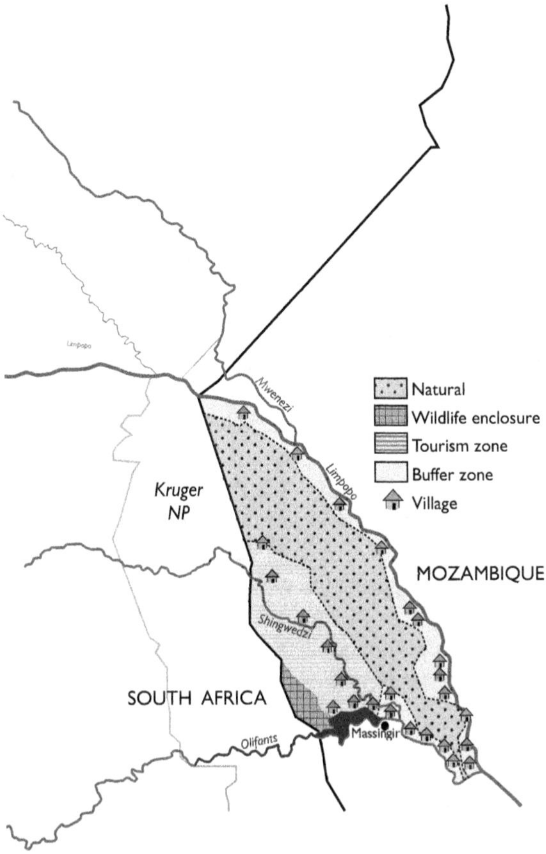

Figure 4.1 The Limpopo National Park showing major settlements and the planned zonation of different land use.

Source: Gazetted Zoning Limpopo National Park, www.limpopopn.gov.mz/maps/gazetted-zoning.aspx (accessed 10 October 2011).

Cross-border migration used to be one of the most dominant forces determining social and household organization during the 1960s in Mozambique, as well as influencing the economy of the region. In some areas as many as three out of four adult men were absent from the household working in South Africa (Tvedten *et al.*, 2010). Labour migration has declined and become informal, turning into a much more insecure livelihood pursuit than when it was regulated. Similar to the Shangaan in Zimbabwe, cross-border migration is seen as a rite of passage for young men (Norman, 2005) and may cost the family as much money as it generates. There is a discrepancy in the number of households who claim to have a member of the family working in South Africa and the number of households who claim to receive remittances. A survey carried out in and adjacent to the LNP ($n = 84$) found that 92 per cent of households had relatives in South Africa, while only 14 per cent of households reported that they received remittances at least occasionally (RRP, 2002). In another survey carried out between 2006 and 2009 ($n = 215$), 34 per cent of households reported remittances as a source of income (Milgroom, 2012).

Between 2001 and 2007, HIV infection rates in Gaza Province, Mozambique, rose from 19 to 27 per cent, and are the highest in the country (World Bank, 2008a). In other provinces, rates have decreased or remained constant over the same time period. The national average prevalence rate is 16 per cent, with some provinces as low as 8 per cent (World Bank, 2008a). The high rates in Gaza are said to be the result of labour migration to South Africa. At a country level, it has been estimated that women aged 20 to 24 are four times more likely to become infected with HIV than men in the same age group. This has important implications for household structure, agricultural productivity and technical knowledge as well as the number of orphaned children. The loss of a male in the household is associated with a major loss of income and a reduction in household consumption (World Bank, 2008b). Men dominate labour migration to South Africa and have a higher potential for wealth accumulation than women who dominate the agriculture sector and informal markets (Tvedten *et al.*, 2010). Some 54 per cent of households in Gaza are currently female-headed and generally worse-off than male-headed households (Tvedten *et al.*, 2010).

Apart from remittances, other sources of income include sale of livestock, sale of agricultural products, transborder trade, sale of locally made or harvested products (such as reed mats, or posts for construction), micro-enterprises, daily wage labour, employment within Mozambique and fishing. The production of charcoal is of primary importance outside of the park, providing an important source of income for 29 per cent of households in villages where charcoal making is permitted (Milgroom, 2012). A study carried out in 2006, a year of drought and food scarcity, observed 100 per cent of households engaging in charcoal production in one form or another as an alternative to agriculture when the cropping season failed (Leonardo, 2007). Formal employment is scarce, and informal employment can be found mainly in construction of houses in the villages or the nearby district capital. Wage labour is found on other people's farms where labour is often paid in maize, or in the charcoal business, where labour is paid mostly in money and often managed by people from outside the area (McGregor, 1998; Brouwer and Magane, 1999). Other labour exchange practices are important for livelihoods, such as the common arrangement for a herder to earn one goat for every year of work with another family's herd. There is no government welfare system and little food aid. Current villages are small (around 100 people on average) and homesteads are dispersed over the landscape, with population density ranging from 2 to 8 people/km^2.

South Africa

Much denser settlements are found on the South African border of the KNP. Marked differences exist from north to south in ethnicity, human density, access to services, socio-economic status and land tenure. These patterns are inextricably linked to South Africa's apartheid past. Discontinuous 'homelands' were demarcated along ethnic lines, resulting in a distinct distribution of ethnic groups; Shangaan and Venda in the north, Shangaan and North Sotho in the central lowveld, and Shangaan and Swazi in the south. These areas continue to be characterized by a high degree of poverty, poor access to services and communal ownership of land, except in some of the emerging urban centres.

Although some people were translocated on creation of the Sabi Game Reserve (the precursor to the KNP) in 1902, this policy was reversed by 1905. Many communities remained in the park until the National Party came into power in 1948 and implemented the apartheid policy (Carruthers, 1995). Until then, local inhabitants in KNP were subjected to strict laws regarding poaching, but were allowed to stay and keep livestock and grow crops. They provided both cheap labour and income for the park since they were tenants on Crown lands who had to pay the government annual rent or provide three months of free labour to the park (Carruthers, 1995). The main removals occurred in the 1950s and 1960s (the Makuleke being the most famous) after the National Party came into power in 1948 and implemented its apartheid policy. However, the dramatic increase in population densities in the 'homelands' on the park boundaries during this period were a result of forced removals and resettlement from across the Transvaal Province, mainly from land earmarked for white ownership for farming. For example, the human population density in the Bushbuckridge area of the central lowveld ranges from 146 people/km^2 to over 300 people/km^2 (High and Shackleton, 2000; Table 4.2). Villages are usually large (over 500 households) and densely packed. Human densities are not as high in the northern communal areas as they are further south, where villages have coalesced in some places.

The resulting high population densities and drastic shortages of land in the 'homelands' forced a transition from an agrarian to a cash-based rural economy (Delius 1996; Gelderblom and Kok, 1994). Seekings (2000) argues that an independent peasantry had been virtually eliminated by the end of the first decade of the apartheid regime. Land-based activities such as agriculture and natural resource use thus play an important safety-net function, but are not a major source of income (Shackleton *et al.*, 2001).

Demographic data collected annually since 1992 in the 24 rural communities of the Agincourt health and socio-demographic surveillance site in the Bushbuckridge local rural municipality of the central lowveld paints a picture of a rural society in rapid transition. The total fertility rate dropped from approximately four children per woman in 1992 to 2.3 in 2004 (Garenne *et al.*, 2007), resulting in a significant decline in average child dependency ratio (number of children 15 years and younger/number between 16 and 59 years) from 0.98 to 0.70 per household. Between 1992 and 2003, mean household size decreased significantly from 6.5 to 5.9. The mortality rates among adults aged 20–49 years increased five-fold over this same period, primarily due to HIV/AIDS (Kahn *et al.*, 2007). HIV prevalence among women at antenatal clinics in Limpopo and Mpumalanga Provinces of South Africa in 2005 was 20 and 33 per cent respectively (Dorrington *et al.*, 2006).

HIV/AIDS and tuberculosis (often associated with HIV) are now the leading causes of death among adults between the ages of 15 and 49 years (Kahn *et al.*, 1999). This has resulted in a near doubling of the number of households with at least one orphan from 2.8 to 5.5 per cent between 1992 and 2003 (Madhavan and Schatz, 2007). Increased mortality among male household heads due to HIV/AIDS may account for the increase in the proportion of female-headed households from 28 to 36 per cent (Madhavan and Schatz, 2007). Households hit by an adult death, especially that of a male wage-earner, often become more reliant on non-timber forest products as a coping strategy, especially to save money and to supplement the household diet (Hunter *et al.*, 2007; Twine and Hunter, 2010). Nevertheless, such households are typically less food secure than non-impacted households, primarily due to the loss of income for buying food and of human labour for growing crops and collecting wild foods.

An influx of mainly Shangaan refugees from Mozambique in the 1970s and 1980s aggravated already high population densities and pressure on natural resources (Kahn *et al.*, 2007). Because they required new residential stands, construction materials and fields to cultivate crops, the arrival of the refugees had a noticeable impact on the local environment (Twine *et al.*, 2003b; Giannecchini *et al.*, 2007). The former Mozambicans who remained in South Africa after the war tended to be poor, with few livestock or sources of income. They are usually highly dependent on non-timber forest products, and are more likely than their South African counterparts to harvest resources such as fuelwood instead of buying them (Giannecchini *et al.*, 2007).

Strong rural–urban linkages are maintained through temporary labour migration. A large proportion of men, and increasing numbers of women, are labour migrants. In Bushbuckridge, 60 per cent of males 35–54 years old are temporary migrants either working or looking for work in urban centres or mines (Collinson, 2009). Social grants, particularly government pensions, are another important income source for poor households in rural South Africa.

Household characteristics such as size, composition and wealth status are important in shaping the role of natural resources in livelihoods of the poor on the edge. For instance, in Bushbuckridge, larger households consume greater quantities of non-timber forest products (Dovie *et al.*, 2005), younger households are less likely to use edible wild herbs and fuelwood (Hansen, 1998), and households with more female members consume more fuelwood (Dovie *et al.*, 2004). Male-headed households are more likely to use bushmeat and wood for carving and house construction (Hansen, 1998). Wealthier households often utilize a wider range of natural resources (Twine *et al.*, 2003a) and use greater amounts of 'luxury' products such as bushmeat, wooden utensils, and construction materials (Twine *et al.*, 2003a). Poorer households consume relatively larger quantities of resources such as edible wild herbs and fruit (Twine *et al.*, 2003a). Households who sell natural products tend to be poor, female-headed households (Shackleton, 2005). Although fewer wealthy households use fuelwood, it remains their main energy source, and quantities consumed by user households are not influenced by wealth status (Dovie *et al.*, 2004).

In the communal areas west of Kruger, the traditional authorities regulate resource harvesting by issuing permits and enforcing laws and taboos. However, due to rapid social, political and economic changes, these institutions have become progressively weakened and marginalized (Twine *et al.*, 2003b; Giannecchinni *et al.*, 2007; Kirkland *et al.*, 2007). This has resulted in increased harvesting pressure in many cases, including felling of live trees, harvesting by 'non-locals', and commercial harvesting of wood and traditional medicine (Twine *et al.*, 2003b; Twine, 2005). Ineffective regulation coupled with high human densities has resulted in shortages of resources such as fuelwood (Giannecchini *et al.*, 2007).

In contrast to the communal areas, the private game reserves, cattle ranches, fruit farms and sugarcane farms west of the KNP in the central lowveld, and on the southern boundary, are still predominantly owned by a white minority. These areas are typically sparsely populated, with relatively affluent owners and management staff, and wage labourers who earn around $180 per month.

Zimbabwe

In Zimbabwe, the colonial era from *c.*1890 to independence in 1980 largely shaped the population and ethnic compositions of people within the GLTFCA. Large numbers of people, especially Ndebele, were moved from expropriated lands in Matabeleland in the west to the Southeast Lowveld (SEL) with the enactment of the Land Apportionment Act of 1930. These people were settled in semi-autonomous villages in the Sengwe Area, predominantly occupied by Shangaan- and Venda-speaking people, with few Hlengwe. Many Karanga moved south to the area from Masvingo after 1930 as a result of state reorganization (centralization) of land into cropping and grazing areas, as well as the abundance of land for cattle keeping and later for cotton production.

Another, albeit small, influx of people onto the edge of what is now Gonarezhou National Park[9] was caused by the establishment of the Gonakudzingwa Native Purchase Area. The Native Purchase Areas – land that could be bought for farming by blacks – were established as early as 1925.[10] These farms were largely occupied by civil servants and businessmen, mostly not native to the area. People were allocated farms on the basis of having acquired agriculture training in the form of a Master Farmer Certificate. The Gonakudzingwa Native Purchase Area formed a buffer between the park and the communal areas and white-owned, large-scale farms further west.

In contrast with Mozambique, where insurgency caused mass migration to South Africa, the Zimbabwe liberation war of the 1970s did not displace large numbers of people from the SEL. Within the region, however, people were forced to live in 'keeps' or 'protected villages' (Logan, 1977), in an attempt to prevent them providing food and cover to Zimbabwe African National Liberation Army (ZANLA) fighters. When people moved out of the keeps they returned to their original homesteads, but other early settlement after the liberation war was haphazard with early settlers preferring forested areas to avoid major roads and settlements. After independence in Zimbabwe, the presence of Mozambican

RENAMO bases in the SEL continued as a disrupting influence on livelihoods, though providing a market for produce and beer-brewing.

Unlike in South Africa where whites still occupy many of the large-scale farms, in Zimbabwe, the recent fast-track land reform has led to the large cattle and wildlife ranches being resettled by black farmers, often with less capital, since 2000. During the chaos resulting from the fast-track land reform of 2000, some people settled within the Gonarezhou National Park boundaries (notably the Chitsa community on the north-east of the park), resulting in an ongoing dispute with the Department of National Parks and Wild Life Management. Population density in the SEL remains much less than that in South Africa (Table 4.2), ranging from around 9 people per km^2 in Sengwe to 34 people per km^2 in Chikombedzi where there is a growth point with a secondary school and hospital.

Although there have been major movements of people from Zimbabwe to South Africa seeking employment for more than a century, the lack of job opportunities in Zimbabwe over the past 10 years has led to increasing numbers of young people, predominantly men, crossing the Limpopo River. This has resulted in large numbers of de facto female-headed households as indicated by the sex ratio of 75 men per 100 women (Table 4.2). Cross-border migration is particularly important for the Shangaan for whom it is a maturity ritual (Scoones and Wolmer, 2003). Recent surveys show that not all households can engage in cross-border migration as it is expensive and requires investment of money for transport, food and bribes along the way. Cross-border migration means that some households become less dependent on agriculture (Murungweni *et al.*, 2011). The remittances allow investment in new assets and technology through remittances in kind of irrigation pumps, solar panels, mobile phones, radios, etc. The gap with poorer households becomes wider as those without remittances remain dependent on agriculture and poorly paid wage labour. This in turn influences the degree of dependence of households on natural resources. A survey in the SEL in 2008 indicated that up to 50 per cent of respondents had no formal education (Murungweni, 2011).

In the Southeast Lowveld of Zimbabwe, sources at Chikombedzi Hospital suggest that the HIV infection rate is above the national average. According to local informants, at the peak of Zimbabwe's economic hardships in 2008 nearly 400 job seekers passed through Chikwarakwara each day, illegally crossing into South Africa. These migrants are also said to have stimulated prostitution, which could partly explain the high HIV/AIDS incidence. HIV/AIDS-affected and child-headed households become economically marginalized with increasing reliance on non-timber forest products, poaching and fishing.

The agricultural economy is largely based on livestock, especially cattle, goats and indigenous chickens and, of late, guinea fowl. Although cattle are important, large numbers are owned by only a few households. Only 20 per cent of households in communal areas of the SEL own ten or more cattle, yet overall 55 per cent of households keep cattle. These other households manage cattle of relatives or friends, and in some instances cattle of absentee cattle owners living in urban areas and South Africa. Cattle rustling to Mozambique is also mentioned as a cause of insecurity by cattle owners, but it is also used as a cover for illegal sale of livestock across the border. Crop production is highly risky due to frequent droughts (locals claim that they only have a good harvest once in every four years). Small grains (mainly sorghum and to a lesser extent millet) predominate though maize is increasingly planted. The shift to maize appears to be due to preference for its taste, because it can be roasted on the cob or eaten as mealie meal, and because it is less susceptible to bird damage and does not need extra labour for scaring birds. Farmers who have access to alluvial soils along the river courses cultivate these fields preferentially as crops are more likely to yield well.

In the SEL, patronage plays different but important roles in the way people accumulate assets such as implements and livestock. Those with political connections have benefited more from land for resettlement and heifer schemes. Access to plots in irrigation schemes, which provides the opportunity to produce vegetables and green mealies for sale, has largely benefited wealthier households (who are also often better politically connected). Households outside the irrigation schemes are more likely to depend on natural resources, e.g., non-timber forest products (mopane worms, Ilala palm, fruits) and poaching during times of drought. On the other hand, relief programmes tend to buffer differences, as well as reducing marginalized households' dependence on natural resources.

Diverse edges around the GLTFCA

Considering only some of the areas where people live within the GLTFCA described above, it is clear that there is great diversity. This is apparent in the timing and stability of settlement on the different edges, the population density, the importance of labour migration and remittances or grants and pensions, and the dependence on crops and livestock. For people living on the edges in all three countries, paid employment in South Africa has always played a major role – though in different ways and to a different extent. In all three areas agriculture, grazing by livestock and harvesting of fuelwood and other non-timber forest products have varying importance depending on other sources of income. We return to these differences later in the chapter after introducing two other important case study areas in Zimbabwe, which illustrate a lesser degree of connectivity with South Africa.

Lower Zambezi-Mana Pools Transfrontier Conservation Area

Until independence in 1980, the Mid-Zambezi Valley was sparsely populated. The Korekore people were the traditional inhabitants, living along the major river valleys that flow north from the escarpment to join the Zambezi River. Formation of Mana Pools National Park in 1963, and the adjacent Sapi and Chewore safari hunting areas in 1964, led to local people being forced to relocate to areas outside the safari areas, although some, who remain predominantly hunter-gatherers, continue to view the safari areas as their hunting grounds. The Mana Pools, Sapi/Chewore complex was designated a World Heritage Site in 1984.

During the liberation war, the valley was an active combat front and the local people were largely forced into protected villages to prevent them providing food to the insurgents. Since independence in 1980, the area of cultivated land has increased exponentially. Cultivation spread initially on the fertile alluvial soils along the major rivers (Baudron *et al.*, 2011). A tsetse eradication programme sponsored by the European Commission in the 1980s made it possible to introduce cattle into the area. This encouraged spontaneous migration into the valley and, in 1986, the government decided to resettle 3,000 families in the area, in addition to the estimated 4,600 resident families, to avert population pressure in areas such as Masvingo. This move dramatically changed the ethnic composition of the area by introducing Karanga-speaking people in the eastern and southern parts of the valley. Schemes to install boreholes on the interfluves that started in 1988 have led to rapid expansion of farming away from the rivers as people became independent of the rivers for drinking water.

Unlike the SEL, there is a strong gradient of increasing intensification of farming with increasing distance away from the safari areas (Table 4.2; Figure 6.1). At first sight, the increase in intensification from the safari areas to Mushumbi Pools appears to be due to early eradication of tsetse fly allowing the influx of cattle. The area under cultivation is determined by the availability of labour, both in terms of animal traction and human labour. But when the available animal traction increases, the cultivated area under cotton tends to increase, while the area under cereal crops remains unchanged. Cotton cultivation has expanded dramatically sponsored by the para-statal Cotton Company of Zimbabwe (CCZ or Cottco) and later by private companies. The amount of income generated by a farm household in the Mid-Zambezi Valley was found to be highly dependent on cotton production (off-farm employment and other remittances do not appear to be a criterion for differentiation; Baudron *et al.*, 2011). Thus there appears to be a positive feedback between cotton and draught animal ownership: cotton generates the initial cash necessary to purchase draught animals, and draught animals provide the traction necessary for expansion of cotton production. Cotton production is therefore the major driver of intensification. On top of this, other factors such as closeness to cotton collection points, tradition of cattle ownership among the Karanga, experience in farming and better education have undoubtedly also played a role in driving rapid expansion of agriculture around Mushumbi Pools.

The fast-track land reform since 2000 has led a new wave of migration into the Zambezi Valley. The requisition of large commercial farms led to the displacement of thousands of farm workers, many of whom were second or third generation descendants of immigrants originally from Malawi or Mozambique. Returning to their ancestral homes in neighbouring countries was not an option for these farm workers and their families who had no local base to seek refuge. The relative abundance of land in the Mid-Zambezi Valley led to many ex-farm workers requesting and being granted land by the local chiefs. An increase in population has also been used by the chiefs to strengthen their power within the areas under their jurisdiction.

By contrast to the areas described above within the GLTFCA, few families in the Mid-Zambezi Valley rely on remittances, and family members working away from home are mainly in Harare. Illiteracy forces many of those who work away to do menial jobs in towns and on the neighbouring resettled former white commercial farms. There are few job opportunities in neighbouring Zambia and cross-border activities are mainly petty trade in cereals, chickens and goats. Rumours indicate substantial trade in wildlife meat and ivory though there is little evidence to substantiate this. Many households harvest wild meat, and forage for fruits and vegetables to supplement their diet, particularly during drought.

Kavango-Zambezi (KAZA) Transfrontier Park

Established in December 2006, the KAZA TFCA is situated in the Okavango and Zambezi river basins where the borders of Angola, Botswana, Namibia, Zambia and Zimbabwe converge. This large (*c.*400,000 km^2) area is comprised of more than 70 different pieces (Table 9.1), representing community conservancies, game management areas, game reserves and national parks, including Bwabwata National Park complex in the Caprivi Strip, Chobe National Park, the Okavango Delta, Hwange and Kafue National Parks, and the Victoria Falls (a World Heritage Site). The KAZA TFCA is host to over 250,000 African elephants, making it the largest contiguous elephant population in the world. The exceptional ecological interest of this sub-region was a major driver of formation of the KAZA TFCA (Ali, 2007). Conservation and tourism are seen by the Peace Parks Foundation to be 'the vehicle for socio-economic development of the region' (PPF, 2009). The area encompassed by the KAZA TFCA is ecologically as well as ethnically, culturally and historically diverse. The area has been influenced by five colonial powers with a range of administration regimes from protectorate to settler colonization that created contrasting historical context (Beinart and Lotte, 2009). The transfer of the Caprivi Strip to Germany at the end of the nineteenth century contributed to the societal and ecological fragmentation of the area. Independence struggles caused considerable disturbance to both human societies and biodiversity (Fritz and Fritz, 1983; Fritz, 1991; Pikirayi, 2001). Post-colonial history is comparatively brief, ranging in period from 10 to 35 years.

We focus on the people living on the edge of the largest protected area of the Zimbabwean component of the KAZA, Hwange National Park, which covers an area of 15,000 km^2. Hwange National Park (HNP) was initially established as a game reserve in 1928. HNP is particularly rich in wildlife due to the diversity of soils and vegetation with more fertile soils mostly on basalt closer to Victoria Falls, and poor soils on the inherently infertile aeolian Kalahari Sands. The Wankie Colliery, a coal mining company on the eastern fringe of the HNP contributed strongly to the economic development of the area since colonial times, probably more that the HNP itself. The Bulawayo–Victoria Falls railway line promoted by the mining activities forms the physical north-east boundary of HNP. Interviews with communities suggest that their displacement when the park was created prevented their access to the more productive vlei-grazing areas. Surface water, as in other semi-arid savannas, is a key constraint of both farmers and wildlife managers. The north-eastern and south-eastern parts of the HNP are naturally dry without surface water and the drilling of waterholes and building of small dams has caused a major change in the ecology of the area. A major consequence has been the conversion of the wet season wildlife ranges into dry season ranges – with much denser wildlife populations and consequently more pressure on interactions between humans and wildlife.

We focus on the Dete area, situated at the northern boundary of HNP close to the Wankie Colliery (Tables 4.1 and 4.2). The case study focused on six villages enclosed between the national park and a Forestry Commission managed Forest. Most of the villages were created in the 1950s when people were resettled at the periphery of the national park. Another village was formed with people resettled from the forest area in 1968. Agriculture is largely for subsistence with small-scale crop and livestock farming. In the traditional Ndebele farming system, most households have two land-holdings: a small cropped field surrounding the homestead and a larger field often at some distance. In the Tsholotsho area to the south-east of HNP, the large fields are often, though not always, on heavier soils and enclosed in high palisades of logs to prevent grazing of crops by livestock and wildlife (Ncube *et al.*, 2009). The area enclosed by the palisade is sometimes shared by different households, and cropping is dominated by sorghum and millet, with smaller areas of groundnut and other crops (Table 4.2). Other major sources of income are employment in the tourism and mining industry, due to the proximity to the main camp of HNP and the Wankie Colliery. In the 1990s, the Dete area relied on a healthy tourism industry based on HNP and benefiting from the Hwange airport and the multiple local safari companies. The economic crisis of the early 2000s followed by US and European sanctions cut down the tourism industry from 95 per cent of occupancy rate in 1999 to 20 per cent in 2009 (90 per cent overseas travellers in 1999 against 10 per cent in 2009). The CAMPFIRE programme that started in the early 1990s contributed to a local development based on wildlife industry, but collapsed in the last decade of crisis, due to a major lack of resources. All these factors induced an increased reliance on subsistence farming to cope with the massive drop in income sources and employment. Most of these villages face increasing land pressure within the communal area, for farming, grazing and natural resource harvesting as the population has grown significantly during the last decade. Harvesting of natural resources for firewood and food products, including wild vegetables, fruits, and mopane worms, constitute crucial resources given the poor agricultural potential of the area. In response to drought in the early 1990s, the Forestry Commission authorized herders to graze their livestock within 3 km from the forest boundary in 1994. Firewood is also collected in the forest, with a special restriction that only women can collect dead wood on Thursdays. Nowadays, the Dete area community is highly dependent on the natural resources provided by the protected areas as a coping strategy in drought years and as a result of the economic decline in Zimbabwe. Strong uncertainties remain on the sustainability of this interface as no buffer area management is clearly established. Water availability remains a crucial issue in the area, but natural resources availability such as grazing areas and wood for building are now of major concern.

Forces that explain population density and shape diverse livelihoods

In this section, we focus on the drivers that have shaped heterogeneity in population density and livelihoods between the study areas on the edge. Although all of the areas are deemed to be marginal in terms of their agro-ecological potential (Table 4.1), population densities vary enormously (Table 4.2). The history of settlement and development in the different areas is very uneven. Relatively dense settlement occurred earlier along the border of KNP in South Africa, in the 1960s and 1970s in the SEL of Zimbabwe, in the 1980s in the Zambezi Valley and since the late 1990s within and around the LNP in Mozambique.

Access to water and land in the flood plains

Of the agro-ecological factors, access to water is a key priority that has controlled the pattern of human settlements in the past, and remains a major factor determining the location of settlements. The land first settled tends to be along the major permanent rivers – the Limpopo, Zambezi and Olifants – and their tributaries. Many local people rely on digging shallow wells in the river beds for water in the dry season. In the SEL, the Mid-Zambezi Valley, and in Mozambique provision of permanent wells and boreholes has relieved the pressure on land close to the rivers and has allowed the spread of settlements, though they tend to remain close to the alluvial soils of flood plains.

Water sources become a major focus of competition between people, livestock and wildlife, particularly during the long dry season. The attraction of the rivers is not due to access of water alone. Riparian vegetation remains green well into the dry season providing preferred forage for both domestic and wild animals. Villages tend to be located along the rivers because of the fertile alluvial soils. The more fertile lands, and lands where irrigation was possible in South Africa, were occupied for commercial farming. Alluvial soils that occur along many of the rivers, sometimes in wide flood plains, form the land with the greatest productive potential for cropping due to the inherent soil fertility. These flat lands are called the *banyeni* in the SEL, or *banhine* in and around the LNP in Mozambique. In general, farmers give greatest priority to planting and tending their crops in these areas in the expectation of good yields, and plant crops on upland soils later. A single flooding episode can allow a crop to be grown to maturity on the residual moisture stored in the soils. Along the permanent rivers, it is possible to irrigate with pumps to provide supplemental irrigation for crops during the main cropping season, and also to produce crops in the dry season. Thus the access to water and concentration of fertile land along the rivers leads them to be a focus of pressure for cropping, livestock and wildlife, which leads to intense competition in these areas. Access to alluvial lands also becomes a key issue for competition between people.

Opportunities for wealth accumulation from agriculture are cotton – which has only played a major role in the Zambezi Valley – and livestock sales. The small irrigated areas close to dams and rivers have benefited a few – but distance to markets for fresh produce limits its importance.

Displacement: war, insecurity, tsetse and natural hazards

Being on the country margins, all of the areas we have considered in this chapter were war zones through which incursions could be made allowing rapid retreat across the border. The relatively later influx of people into the Zambezi Valley and in and around the area that is now the LNP in Mozambique was, at least partly, due to the fact that these areas were strongly impacted by war. Both were centres of active engagement between the warring parties. In Zimbabwe, the local population was forced into 'protected villages' and the activities of the Rhodesian army prevented substantial settlement in the Mid-Zambezi Valley. An additional factor contributing to the later settlement of the Zambezi Valley was the prevalence of tsetse infestation – although attempts to clear tsetse from the valley go back to the 1950s these were interrupted by the war, which led to resurgence of the fly. Tsetse was actively combated in the SEL of Zimbabwe presumably because of the nearby commercial livestock enterprises.

The area of the LNP was a major source of conflict between RENAMO and FRELIMO and much of the population was relocated to FRELIMO villages. This resulted in major cross-border movements into South Africa of people fleeing the civil war in Mozambique. Further displacement of settlements along the Limpopo has occurred due to extreme flooding events. Although forced removal of people occurred upon the creation of the Kruger National Park, and a resettlement programme is ongoing in the LNP, overall the establishment and presence of the parks and conservation areas does not seem to play a major role in determining population densities on their edges.

Rural–urban connections and (cross-border) migration

Labour migration, rural–urban connections and more specifically cross-border migration are important in shaping livelihoods and heterogeneity in the edges. Resulting in skewed sex distributions and relatively lower population densities because of absenteeism, urban-derived remittances are a major livelihood source for rural households in the GLTFCA areas of all three countries, and in the KAZA TFCA area of Zimbabwe. In Mozambique and southern Zimbabwe, more than a century of migration to South Africa has led to this becoming institutionalized as evidence of manhood, and remittances remain an important source of income. By contrast, in the Mid-Zambezi Valley, migration for work is mainly focused on Harare, with the majority of families relying mainly on agriculture for their income. Cross-border movements appear mainly to be for trade rather than for paid work. Migration is also one factor often linked to the incidence of HIV/AIDS.

Prevalence and impacts of HIV/AIDS

Although accurate comparative statistics on HIV/AIDS are lacking, it appears the high prevalence on the edge is related to high levels of migration and mobility. Many men, and increasing numbers of women, work on a semi-permanent basis in South Africa, often returning to visit their family homesteads only once a year. Further, the border areas with South Africa are also important routes for migrants coming from other parts of Mozambique and Zimbabwe, or even further afield from countries such as Malawi. The greater incidence of HIV/AIDS brings problems of single (female)-headed, and child-headed households, loss of income and increased vulnerability, but it does not appear to be a major factor explaining population densities on the edge.

Segregationist land policies and resettlement

By far the most important drivers explaining current differences of population densities on the edges are the segregationist land policies and resettlement. In South Africa, forced removals into the black 'homelands' resulted in population densities of more than 300 people/km^2, far more than can be supported by the land. Expropriation of land in Zimbabwe during the colonial era led to waves of settlement into the SEL. In the Mid-Zambezi Valley, the population density has increased dramatically since independence in 1980 with multiple factors influencing resettlement. This was initially driven by government-planned resettlement that triggered further influx. The clearing of tsetse fly made much of the area more attractive for farming as livestock could be used for ploughing, but cotton cultivation has been a major driver of clearance of land for agriculture (Baudron *et al.*, 2011). Since 2000, a new wave of settlement occurred in the Zambezi Valley as farm workers displaced from commercial farms sought new opportunities. Thus here, and in other areas, spontaneous migration has occurred due to the relative abundance of land, despite the remoteness from urban centres and the relatively poor rainfall. Thus across all areas we can understand population growth as the product of both planned and spontaneous settlement.

Emerging types of livelihoods

Despite all the differences we find across the countries and locations within countries, we can recognize similar patterns of livelihoods that are often found (see for example, Murungweni *et al.*, 2011). Understanding the different types of livelihoods, and the forces that shape them, may form the basis for common interventions across the region addressing similar impacts and problems.

Livelihoods are comprised of different pursuits, the relative importance of which varies enormously. Thus we can distinguish livelihoods that are:

1 dependent mainly on cropping;
2 dependent mainly on keeping cattle;
3 primarily dependent on other forms of local paid work, hiring out their labour or from hunting or harvesting of other non-timber forest products;
4 dependent mainly on external income from remittances from household members in paid employment, or in South Africa on income from welfare/child grants or pensions.

These are, of course, gross generalizations and many households encountered do not fall comfortably into one category, but tend to blend characteristics of more than one type. Nevertheless, identifying the diversity of types of livelihoods on the basis of their major pursuits provides insights for understanding their likely interest in, and response to, new policies or development interventions. At least it takes us one step further in recognizing and attempting to engage and understand the diversity of livelihood approaches rather than treating local people as a homogeneous group of 'smallholder farmers'. As we have seen their livelihoods may often hold farming central, but are dependent on a wide diversity of resources and forms of income – even for their food security. Looking across the different countries and locations described in this chapter, it appears that when more land is available (i.e. less people) livelihoods are more dependent on natural resources and when there are more people, they are more dependent on external employment and pensions.

Conclusions

To return to the idea that conservation areas and parks act as 'honey pots' that draw people towards them due to the livelihood opportunities they offer (Wittemeyer *et al.*, 2008), we find little evidence for such an effect. Indeed, Joppa *et al.* (2009) demonstrated that the conclusions of Wittemeyer *et al.* (2008) were an artefact due to comparison of incompatible datasets. Nevertheless, an equal number of the cases analysed by Joppa *et al.* (2009) showed an increased population as the number that showed a decrease in population close to parks. We set out to understand the drivers of population growth on the edge, and find, perhaps unsurprisingly that almost all drivers played a role everywhere, yet to varying degrees. The major drivers we identify are: (1) displacement through war and insecurity; (2) labour reserves and (re)settlement policy; (3) rural–urban connections and labour migration; and (4) the opportunity offered by the (relative) abundance of land along the edges of conservation areas. It is difficult, if not impossible, to unravel the relative importance of these drivers in every case. Changes in population are gradual, and many drivers act together and with synergy, or sometimes antagonistically. But perhaps the major difficulty in explaining population change is that the most important driver is indirect – the (lack of) alternative opportunities away from the edge. Rapid population growth at national level, combined with poor economic growth (and even economic contraction and de-industrialization in Zimbabwe) led to people seeking places to settle. The remoteness and relatively poor agro-ecological potential of the conservation areas and environs means that, while they may be the ancestral home for some, they are not the first choice for in-migration and settlement. In Sengwe, Southeast Lowveld, Zimbabwe, the lack of trust that local people have in the development of the TFCA actually appears to be a 'push' factor triggering out-migration because of the uncertainty that it has created.

Past resettlement policy appears to be the overriding factor that has influenced population density close to the conservation areas examined in Southern Africa, but superimposed on this are many other factors. Focusing more closely on the local level, we can attempt to unravel the granularity of these influences. The forces that stimulate migration towards the national parks seem to be more the availability of land for activities such as food production, livestock grazing and cotton-growing (for example in the Mid-Zambezi Valley), rather than benefits of 'living on the edge' through opportunities offered by the conservation areas themselves such as resources or eco-tourism. Equally, there are forces that draw people away from the conservation areas. Two factors – the proximity of the national borders with Botswana or South Africa, and the allure and security of paid employment relative to the lack of local opportunity – act as a magnet pulling men, and the youth, away from the edge. The 'remoteness' of the edges has offered opportunities for 'border hopping' that were sufficiently attractive to risk the (often fatal) danger of attack by predators. Yet the rural home with its ancestral ties remains a place to live and belong although it may not be the primary source of income or livelihood. The opportunity remains to access further land through kinship for young families. The net result is more population on the edge, even when the men are pulled away from it.

Understanding diversity on the edge helps to uncover the underlying factors that often result in increased vulnerability of poorer households. We demonstrate that there is great heterogeneity between the livelihoods of households in the areas close to the edge of the parks. Heterogeneity is shaped and characterized by socio-economic differentiation such as ethnicity (see Chapter 5), education, farming practices, technologies employed, natural resource use (including land) and income sources, many of which are interdependent. This heterogeneity is demonstrated at a range of scales – between countries, between regions within countries, between land use types (small-scale commercial agriculture *vs.* communal agriculture), between villages (depending on proximity to the edge), between households within villages. Further attributes of households, such as size and gender, remittances, access to land of different quality, livestock ownership and use and cropping patterns also vary widely, demanding a comparative approach to understand their importance.

Any policy or intervention is likely to have strongly differential effects on different households, both in terms of the interpretation and understanding of the policy or intervention, and in the impacts. Some of the impacts will be unexpected or unintended, because of the range in capacity of the households to respond. Targeting is sometimes attempted, for example food relief is generally intended for the most needy areas and needy households, but is delivered to them with varying success. This raises the question: is it possible to target policies and interventions to the specific types of conditions created by living on the edge? This question will be returned to later in this volume when we consider the future for both conservation and people living on the edge.

Acknowledgements

We thank Billy Mukamuri, Petronella Chaminuka, Chaka Chirozva, Collen Matema, Xavier Poshiwa, Tendai Nzuma for providing information and for comments and suggestions on the text. We are grateful to Kefasi Nyikahadzoi for his review.

Notes

1 State-protected areas cover nearly 12 per cent of Southern Africa.
2 www.pifworld.com/#/projects/theelephantcorridor/61 (accessed 29 July 2011).
3 GLTFP: www.greatlimpopopark.com/.
4 KAZA: www.kazapark.com/.
5 Lower Zambezi-Mana Pools Transfrontier Conservation Area: www.awf.org/content/heartland/detail/1285.
6 The term Shangaan, commonly used to describe the residents of the region, is derived from Shoshangane, the first king of the Gaza Nguni, but does not accurately denote ethnicity (Junod, 1962; Liesegang, 1977; Harries, 1989).
7 The number of recruits peaked at 13,016 in 1942 – 8,996 from Pafuri and 4,020 from the East Coast (WNLA, 1943 Salisbury file: 46B Pad1).
8 With funding from KfW (the German Development Bank, Kreditanstalt für Wiederaufbau).
9 Although Gonarezhou Game Reserve was first proclaimed in the area in 1928, this was later deproclaimed and then reinstated. Malipati Game Reserve was established in 1968, but was later redesignated as a Safari Area in 1975, the same time as an expanded Gonarezhou National Park was established (Murphree and Cumming 1993). See also Chapter 3.
10 The Native Purchase Areas were initially 182 km^2 in extent but was increased to 30,190 km^2 in 1931. This had been reduced to 17,078 km^2 by 1964, largely because the farms were not taken up and the remaining area was converted to Native Reserve, and later Tribal Trust Land (now known as Communal Areas). After independence Native Purchase Areas became known as small-scale commercial farms.

References

Ali, S.H. (2007) *Peace Parks: Conservation and Conflict Resolution*, MIT Press, Cambridge, MA

Andersson, J.A. (2006) 'Informal moves, informal markets: international migrants and traders from Mzimba District, Malawi', *African Affairs*, vol. 105, no. 420, pp. 375–397

Baudron, F., Corbeels, M., Andersson, J.A., Sibanda, M. and Giller, K.E. (2011) 'Delineating the drivers of waning wildlife habitat: the predominance of cotton farming on the fringe of protected areas in the Mid-Zambezi Valley, Zimbabwe', *Biological Conservation*, vol. 144, pp. 1481–1493

Beinart, W. and Lotte, H. (2009) *Environment and Empire*, Oxford University Press, Oxford

Brouwer, R. and Magane, D. (1999) 'The charcoal commodity chain in Maputo: access and sustainability', *Southern Africa Forestry*, vol. 185, pp. 27–34

Carruthers, J. (1995) *The Kruger National Park: A Social and Political History*, University of Natal Press, Pietermaritzburg

Collinson, M.A. (2009) 'Striving against adversity: the dynamics of migration, health and poverty in rural South Africa', Ph.D. thesis, Umea University, Umea

Cumming, D.H.M. and Lynam, T.J.P. (1997) *Landuse Changes, Wildlife Conservation and Utilisation, and the Sustainability of Agro-ecosystems in the Zambezi Valley*, WWF, Harare

Davies, R. and Head, J. (1995) 'The future of mine migrancy in the context of broader trends in migration in Southern Africa', *Journal of Southern African Studies*, vol. 21, no. 3, pp. 439–450

Delius, P. (1996) *A Lion Amongst the Cattle: Reconstruction and Resistance in the Northern Transvaal*, Raven Press, Johannesburg

Dorrington, R., Johnson, L.F., Bradshaw, D. and Daniel, T. (2006) *The Demographic Impact of HIV/AIDS in South Africa. National and Provincial Indicators for 2006*, Centre for Actuarial Research, South African Medical Research Council and Actuarial Society of South Africa, Cape Town

Dovie, D.B.K., Witkowski, E.T.F. and Shackleton, C.M. (2004) 'The fuelwood crisis in Southern Africa – relating fuelwood use to livelihoods in a rural village', *GeoJournal*, vol. 60, pp. 123–133

Dovie, D.B.K., Witkowski, E.T.F. and Shackleton, C.M. (2005) 'Monetary valuation of livelihoods for understanding the composition and complexity of rural households', *Agriculture and Human Values*, vol. 22, pp. 87–103

Dzingirai, V. (2004) *Disenfranchisement at Large: Transfrontier Zones, Conservation and Local Livelihoods*, IUCN-ROSA, Harare

Fritz, J-C. (1991) *La Namibie Indépendante: Le Coût d'Une Décolonisation Retardée*, L'Harmattan, Paris

Fritz, J-C. and Fritz, G. (1983) *Le Zimbabwe Indépendant. Décolonisation et Stratégie de Développement. Annuaire du Tiers-monde*, no. VII, 1980/1981, Istra, Paris

Garenne, M.L., Tollman, S.M., Collinson, M. and Kahn, K. (2007) 'Fertility trends and net reproduction in Agincourt, rural South Africa: 1992–2004', *Scandinavian Journal of Public Health*, vol. 35, pp. 68–76

Gelderblom, D. and Kok, P. (1994) *Urbanization: South Africa's challenge. Volume 1: Dynamics*, HSRC Publishers, Pretoria

Giannecchini, M., Twine, W. and Vogel, C. (2007) 'Land-cover change and human–environment interactions in a rural cultural landscape in South Africa', *The Geographical Journal*, vol. 173, pp. 26–42

Hansen, B. (1998) 'Changing patterns of natural woodland resource dependency and use: intergenerational perceptions, traditions and customs', MSc thesis, University of Copenhagen, Copenhagen

Harries, P. (1989) 'Exclusion, classification and internal colonialism: the emergence of ethnicity among the Tsonga speakers of South Africa', in L. Vail (ed.), *The Creation of Tribalism in Southern Africa*, University of California Press, Berkeley, pp. 82–117

Harries, P. (1994) *Work, Culture and Identity: Migrant Labourers in Mozambique and South Africa c. 1860–1910*, James Currey, London

High, C. and Shackleton, C.M. (2000) 'The comparative value of wild and domestic plants in home gardens of a South African rural village', *Agroforestry Systems*, vol. 48, pp. 141–156

Hunter, L.M., Twine, W. and Patterson, L. (2007) '"Locusts are now our beef": adult mortality and household dietary use of local environmental resources in rural South Africa', *Scandinavian Journal of Public Health. Supplement*, vol. 35, pp. 165–174

Joppa, L., Loarie, S. and Pimm, S. (2009) 'On population growth near protected areas', *PLoS ONE*, vol. 4, e4279

Junod, H. (1962) *Life of a South African Tribe. Social Life, Part 1*, University Books, New Hyde Park, New York.

Kahn, K., Garenne, M.L., Collinson, M.A. and Tollman, S.M. (2007) 'Research into health, population and social transitions in rural South Africa: data and methods of the Agincourt health and demographic surveillance system', *Scandinavian Journal of Public Health*, vol. 35, pp. 8–20

Kahn, K., Tollman, S.M., Garenne, M. and Gear, J.S.S. (1999) 'Who dies from what? Determining cause of death in Africa's rural north-east', *Tropical Medicine and International Health*, vol. 4, pp. 433–441

Katerere, Y., Hill, R. and Moyo, S. (2001) *A Critique of Transboundary Natural Resource Management in Southern Africa*, the IUCN-ROSA series on Transboundary Natural Resource Management, IUCN, Regional Office for Southern Africa, Harare

Kirkland, T., Hunter, L. and Twine, W. (2007) '"The Bush is no more": insights on institutional change and natural resource availability in rural South Africa', *Society and Natural Resources*, vol. 20, pp. 337–350

Leonardo, W. (2007) 'Patterns of nutrient allocation and management in smallholder farming systems in Massingir District, Mozambique. A case study of Banga Village', MSc thesis, Wageningen University, Wageningen

Liesegang, G. (1977) 'New light on Venda traditions: Mahumane's account of 1730', *History in Africa*, vol. 4, pp. 163–181

Liesegang, G. (1983) 'A first look at the import and export trade of Mozambique, 1800–1914', in G. Liesegang, H. Pash and A. Jones (eds), *Symposium on the Quantification and Structure of the Import and Export and Long Distance Trade of Africa in the 19th Century (c. 1800–1913)*, St Augustin, FL, pp. 452–523

Logan, D. (1977) 'Rhodesia – the dilemma of the man in the middle', *The Third Way: Towards a Biblical World View*, vol. 1, pp. 6–9

Lunstrum, E. (2008) 'Mozambique, neoliberal land reform and the Limpopo National Park', *Geographical Review*, vol. 98, p. 339

McGregor, J. (1998) 'Violence and social change in a border economy: war in the Maputo Hinterland, 1984–1992', *Journal of Southern African Studies*, vol. 24, pp. 37–60

Madhavan, S. and Schatz, E.J. (2007) 'Coping with change: household structure and composition in rural South Africa, 1992–2003', *Scandinavian Journal of Public Health*, vol. 35, pp. 85–93

Mandlate, C. and Machele, J. (2009) *Historia das comunidades de Macavene e Nanguene*, unpublished report, Universidade Eduardo Mondlane, Maputo

Mavhunga, C. and Spierenburg, M. (2007) 'A finger on the pulse of the fly: hidden voices of colonial anti-tsetse science on the Rhodesian and Mozambican borderlands 1945–1956', *South African Historical Journal*, vol. 58, pp. 117–141

Milgroom, J. (2012) 'The elephants of democracy: an unfolding process of resettlement in the Limpopo National Park', Ph.D. thesis, Wageningen University, Wageningen

Milgroom, J. and Spierenburg, M. (2008) 'Induced volition: resettlement from the Limpopo National Park, Mozambique', *Journal of Contemporary African Studies*, vol. 26, pp. 435–448

Murphree, M.W. and Cumming, D.H.M. (1993) 'Savanna land use: policy and practice in Zimbabwe', in M.D. Young and O.T. Solbrig (eds), *The World's Savannas: Economic Driving Forces, Ecological Constraints and Policy Options for Sustainable Land Use*, Man and the Biosphere Series, UNESCO, Paris, and The Parthenon Publishing Group, London, pp. 139–178

Murungweni, C. (2011) 'Vulnerability and resilience of competing land-based livelihoods in South Eastern Zimbabwe', Ph.D. thesis, Wageningen University, Wageningen

Murungweni, C., van Wijk, M.T., Andersson, J.A., Smaling, E.M.A. and Giller, K.E. (2011) 'Application of fuzzy cognitive mapping in livelihood vulnerability analysis', *Ecology and Society*, vol. 16, no. 4: 8. Available online: www.ecologyandsociety.org/vol16/iss4/art8/

Ncube, B., Twomlow, S., Dimes, J., van Wijk, M.T. and Giller, K.E. (2009) 'Resource flows, crops and soil fertility management in smallholder farming systems in semi-arid Zimbabwe', *Soil Use and Management*, vol. 25, pp. 78–90

Newitt, M. (1995) *A History of Mozambique*, Hurst and Co, London

Norman, W.O. (2005) 'Living on the frontline: politics, migration and transfrontier conservation in the Mozambican villages of the Mozambique-South Africa Borderland', Ph.D. thesis, London School of Economics and Political Science, London

Pikirayi, I. (2001) *The Zimbabwe Culture: Origins and Decline of Southern Zambezian States*, AltaMira Press, Walnut Creek, CA

PPF (2009) *Kavango-Zambezi Transfrontier Conservation Area, Integrated Development Plan, Zimbabwe Component.* ZPWMA/Peace Parks Foundation, Harare

Ranger, T. (1979) 'The mobilization of labour and the production of knowledge: the antiquarian tradition in Rhodesia', *Journal of African History*, vol. 20, pp. 507–524

RRP (2002) *A Park for the People? Great Limpopo Transfrontier Park – Community Consultation in Coutada 16, Mozambique*, Refugee Research Programme, University of Witwatersrand, Johannesburg

Scoones, I. and Wolmer, W. (2003) 'Introduction: livelihoods in crisis: challenges for rural development in Southern Africa', *IDS Bulletin*, vol. 34, pp. 1–11

Seekings, J. (2000) 'Visions of society: peasants, workers and the unemployed in a changing South Africa', *Journal for Studies in Economics and Econometrics*, vol. 24, pp. 53–72

Shackleton, S.E. (2005) 'The significance of the local trade in natural resource products for livelihoods and poverty alleviation in South Africa', Ph.D. thesis, Rhodes University, Grahamstown

Shackleton, C.M., Cousins, B. and Shackleton, S.E. (2001) 'The role of land-based strategies in rural livelihoods: the contribution of arable production, animal husbandry and natural resource harvesting in communal areas in South Africa', *Development Southern Africa*, vol. 18, pp. 582–604

Smith, A. (1973) 'The peoples of southern Mozambique: a historical survey', *Journal of African History*, vol. 14, pp. 565–580

Tvedten, I., Paulo, M. and Tuominen, M. (2010) '"A woman should not be the boss when a man is present": gender and poverty in Southern Mozambique', CMR Report R2010:7, Chr. Michelsen Institute, Bergen

Twine, W. (2005) 'Changing socio-economic factors influence vegetation change in the communal rangelands of the South African Lowveld', *African Journal of Range and Forage Science*, vol. 22, pp. 93–100

Twine, W. and Hunter, L.M. (2010) 'Adult mortality, food security and the use of wild natural resources in a rural district of South Africa', in A. Niehof, G. Rugalema and S. Gillespie (eds), *AIDS and Rural Livelihoods: Dynamics and Diversity in Sub-Saharan Africa*, Earthscan, London, pp. 153–170

Twine, W., Moshe, D., Netshiluvhi, T. and Siphugu, V. (2003a) 'Consumption and direct-use values of savanna bio-resources used by rural households in Mametja, a semi-arid area of Limpopo province, South Africa', *South African Journal of Science*, vol. 99, pp. 467–473

Twine, W., Siphugu, V. and Moshe, D. (2003b) 'Harvesting of communal resources by "outsiders" in rural South Africa: a case of xenophobia or a real threat to sustainability?', *International Journal of Sustainable Development and World Ecology*, vol. 10, pp. 263–274

Wittemeyer, G., Elsen, P., Bean, W.T., Burton, A.C.O. and Brashares, J.S. (2008) 'Accelerated human population growth at protected area edges', *Science*, vol. 321, pp. 123–126

Witter, R. (2010) 'Taking their territory with them when they go: mobility and access in Mozambique's Limpopo National Park', Ph.D. thesis, University of Georgia, Athens, GA

World Bank (2008a) *Mozambique: HIV/AIDS and Nutrition Status Report*, World Bank, Washington DC

World Bank (2008b) *Beating the Odds: Sustaining Inclusion in a Growing Economy: A Mozambique Poverty, Gender, and Social Assessment*, Report No. 40048-MZ, World Bank, Washington DC

5 Ethnic heterogeneity and its implications for natural resources management on the edge

Billy Mukamuri, Chaka Chirozva, Collen Matema, Steven Matema and Tendai Nzuma

Introduction

However ill-defined transfrontier conservation areas (TFCAs) remain, the underpinning idea that conservation should benefit local peoples and contribute to development goes hand in hand with the assumption that natural resources in these areas are collectively managed. While the previous chapter discussed the drivers of population density on the edge of protected wildlife areas and the diversity of livelihoods in similar agro-ecological circumstances, this chapter explores the consequences of such a diverse population on the edge for collective natural resource management. It examines the commonly used notion of 'community' in natural resource management from a socio-cultural angle, by developing a perspective on the ethnic configurations of people living within TFCAs.

Studies of community-based natural resource management (CBNRM) have revealed that the notion of 'community' and its related assumption of socio-cultural homogeneity are highly problematic. Community-based projects often fail because a 'collective' appears to be absent; suspicion among locals, local political conflicts, lack of 'community' participation and lack of interest among locals prevent collective action (Songorwa, 1999; Alexander and McGregor, 2000; Mukamuri *et al.*, 2000; Logan and Moseley, 2002; Musumali *et al.*, 2007; Balint and Mashinya, 2008; Mutimukuru-Maravanyika, 2010). Such local-level dynamics and struggles are often expressed in ethnic terms, despite the fact that ethnic distinctions are first and foremost socio-historical constructions, rather than ontological categories.

Two uses of ethnicity may be distinguished, both referring to different academic understandings of the term. First, ethnicity may be taken as a marker of different livelihood pursuits and resource use among local peoples, and signify disparate attitudes towards wildlife and other natural resources in TFCAs. In such use, ethnicity may appear as an ontological category. Second, ethnic distinctions may be used strategically. For instance, local people may deploy an ethnic discourse to claim resource access and use rights (for instance, by claiming ethnic seniority), while outsiders such as government agencies and non-governmental organizations

(NGOs) may construct and use (or ignore completely) ethnic classifications to pursue particular policies.

Navigating between these seemingly contradictory understandings of ethnicity as a marker of existing difference and a resource for creating difference, this chapter argues that ethnic heterogeneity is an important, and potentially disrupting, factor in TFCA governance. As TFCA development progresses with landscape plans made at higher policy levels, the relevance of ethnic distinctions for the management of such landscapes tends to be ignored. Consequently, the basis for collective natural resource management in TFCAs is undermined at the outset.

The chapter first briefly explores theoretical understandings of ethnicity and ethnic boundaries and their relevance for natural resource management issues. Second, it discusses how ethnic distinctions gained relevance during the colonial era in areas on the fringe of protected areas in Southern Africa. It is shown how colonial governments reconfigured ethnicity by moving large numbers of people to distant areas already inhabited by other groups. Third, we argue for recognition of ethnic diversity by natural resource planners working in areas adjacent to protected areas.

Theoretical aspects of ethnicity

A reading of Barth (1966), an anthropologist, sheds light into what ethnicity is and what ethnic groups are. An ethnic group or ethnicity is a sub-sector of a broader culture which separates groups of people from one another. Ethnicity, therefore, represents discrete patterns of behaviour that describe the specific group of people. The existence of ethnicity does not necessarily correspond to a space nor is it limited to social segregation, as many people would like to believe. Ethnicity can continue to exist even if different groups do not live in the same geographical and social spaces (Barth, 1966). Ethnicity serves to ascribe social identity. Ethnicity also perpetuates itself through a number of processes such as procreation, value sharing, communication and identity, the latter also shaped by where the person originated from and other socio-historical factors. When living in a heterogeneous community, aspects such as language, lifestyle and dress can be used as ethnic bound symbols. More importantly, people who identify themselves as members of distinct ethnicity tend to share rules and these may be extended to cover activities they engage in (Barth, 1966). Above all, ethnicity has the potential to create bonds among people, largely marked by solidarity and organization.

Further reading of Barth (1966) shows that ethnicity can also be applied in the realm of managing natural resources. For example, these ethnic groups can exclusively utilize specific ecological areas (niches), thereby monopolizing certain natural resources. In cases where other ethnic groups live in the same area and are interested in utilizing the same resource, competition and even conflict may arise.

Demographic characteristics also play an important role in determining access to particular resources such as markets and environmental goods. The real issue emerges when one or more distinct ethnic groups have to utilize a common resource. This then calls for either the different groups to negotiate on limiting their use of the resource, or if not, then competition and conflicts emerge. Once the latter happens, the resource in question is affected. We would also assume that when such a situation emerges, out-migration can play an important role in reducing pressure on resources. According to Barth, when thinking about resource utilization and management in what he calls a polyethnic society, we also need to understand the complexity that underlies the whole socio-ecological system.

Ethnicity can also define social stratification, particularly when an ethnic group dominates the means of production or is naturally inclined to a more economically lucrative activity. Stratification or rather socio-economic differentiation is likely to lead to ethnic reinforcement, as a strategy to continue dominating that particular livelihood route. For example, stratification and ethnicity have managed to maintain caste systems in India and in Ethiopia, where they have been used to influence how different ethnic groups are related to the state (Barth, 1966).

We are also reminded how ethnicity is influential in the making of social groups that are largely marginal or peripheral to the system. Here we are reminded of minorities and pariahs. These characteristics can be created by both internal and external dynamics, also shaped by historical processes (Barth, 1966). A social system with these attributes can be characterized by vices such as exclusion in terms of access to decision-making and utilization of ecological resources. In general, weak minorities and pariahs face difficulties in meeting their needs and are largely forced to be incorporated into the dominant ethnic groups where they face challenges such as discrimination and exclusion. For these groups to fully benefit from the polyethnic system, they need to be highly innovative and change some of their social attributes to fit into the bigger model.

Ethnic groups across the world are conscious of the fact that present-day boundaries are products of political and state administrative structures, hence the perception that the boundaries are artificial (Powell, 2000; Hazarika, 2004). Ethnicity can be invoked to facilitate access to resources and power (Wallace, 2003; O'Sullivan and Dana, 2008; Lee, 2009). Ethnicity is also widely believed to be a product of a continuous process, usually engendering and embedded in contestation (Niehaus, 2002; Wallace, 2003; Virtanen, 2005; Lee, 2009). Appeals to ethnicity are not only for cultural reverence and preservation, but are also used as a tool for exclusion and inclusion. Through time therefore, ethnic identity is constructed, deconstructed and reconstructed within a broader political ecology (O'Sullivan and Dana, 2008).

The construction and entrenchment of ethnic identities does not take place in a vacuum. Ford and Harawa (2010) argue that ethnicity is complex and socially constructed with implications for societal interactions with spatial and temporal dimensions. In general ethnicity is used to refer to individuals sharing common attributes that differentiate them from other social groups and entities within a given society (Marshall, 1998; Ford and Harawa, 2010). Ethnic groups are an outcome of descent-based attributes or acquired identity for membership (Chandra, 2011) and this has a direct relationship to resource access and management practices (O'Sullivan and Dana, 2008). In Southern Africa, ethnicity has been mainly studied in relation to the influence of the colonial state or the independent nation-state: either as 'invented' or 'imagined' (Ranger, 1983, 1985a, 1999) or as a result of state colonial or post-colonial violence, which resulted in forced displacements of certain ethnic groups from the central plateau to the Southeast Lowveld (Werbner, 1995).

Specific to natural resources use, wildlife included, Ostrom (2007, 2009) identifies cultural factors, of which ethnicity is one, that influence collective action, believed to be key in bringing about long-term sustainable management of natural resources. This chapter deals with the ethnic configurations of people living within TFCAs. We also explore ethnic diversity, particularly its implications for natural resource conservation, collective action and governance. We use this exploratory presentation to urge policy makers strongly that they should avoid generating panaceas by using a 'one-size-fits-all' approach when dealing with complex ethnic, environmental and natural resource management issues. The argument we put forward is that ethnic diversity sometimes leads to disparate attitudes towards wildlife and this has negative implications for wildlife-based development activities in the proposed transfrontier areas. We direct this advice especially to development agents and organizations working in transfrontier wildlife conservation areas of Southern Africa (South Africa, Zimbabwe and Mozambique to a lesser extent). The selected countries are important because they are pioneering one of the largest TFCAs in Southern Africa.

Ethnicity and CBNRM in Southern Africa

Studies conducted in Southern Africa, focusing on attempts to implement CBNRM initiatives, indicate that people living in rural areas are heterogeneous and their actions are influenced by ethnicity, origin and other socio-economic variables such as cattle ownership, education and religion (Mukamuri, 1989; Dzingirai, 1996, 2003; Mukamuri *et al.*, 2000; Mutimukuru-Maravanyika, 2010). Historical factors such as pre-colonial migrations, colonial relocation of Africans to marginal areas, civil wars and land reforms are arguably some of the major drivers of community heterogeneity in Southern Africa (Ranger, 1985a; Phimister, 1988). Colonial-induced migrations have been reported by Bocchino and Murphree (2010) to have resulted in the formation of distinct cultural districts.

Colonial administrative policies characterized by pacification of rebel community leaders and promotion of pro-state leaders contributed to the legitimacy crisis currently affecting local level leadership structures (Niehaus, 2002; Ranger, 1985b). Post-colonial governments, particularly in Tanzania, Mozambique and Zimbabwe, further complicated the issue by introducing new leadership structures seeking to undermine traditional leaders in favour of quasi-state institutions meant to be accountable to the state (Murombedzi, 1992). Similar strategies were employed in South Africa where traditional leaders who supported the apartheid regime were promoted and appointed leaders of the Bantustans (Niehaus, 2002; Connor, 2003). Policies and practices by different governments during the last century, largely aimed at manipulating local leaders for conquest and political gains, resulted in weak local leadership structures in most areas, including those at the edge of most national parks in Southern Africa (Murombedzi, 1992; Dzingirai, 1996).

In short, we now remain with weak leadership structures incapable of coordinating functionally different ethnic units for both natural resource management and an array of other community-based development initiatives, CAMPFIRE included (Dzingirai, 1996; Murombedzi, 1999; Murphree, 2001). Literature abounds with examples of community projects that failed because of suspicion among locals, local political conflicts, lack of community participation and lack of interest among locals (Songorwa, 1999; Alexander and McGregor, 2000; Mukamuri *et al.*, 2000; Logan and Moseley, 2002; Musumali *et al.*, 2007; Balint and Mashinya, 2008; Mutimukuru-Maravanyika, 2010).

Southern Africa is also characterized by a history of relocation of indigenous people to agriculturally marginal lands, such as most of the areas adjacent to national parks. This was a result of the grabbing of large tracts of land, which started when European settlers failed to find another Eldorado with massive deposits of gold to sustain their commercial interests. In Mozambique, the Portuguese introduced large commercial farms known as *prazos* at the turn of the nineteenth century. Although the colonial settlements in Zimbabwe began during the 1890s, large-scale forced movement of Africans started in the late 1930s following the enactment of the Land Apportionment Act (1930). In most cases, translocations involved whole clans, villages and chiefdoms. This process is particularly elaborate on the Zimbabwean edge, especially in the north-western and south-eastern parts of the country. Translocation of people and their leadership structures into areas already presided over by leaders of different ethnic origins created conflicts and animosity among people, as well as a confusing overlay of authority. Common across the three countries was the annexation of the most fertile land with Africans being moved to marginal lands. In Zimbabwe and Mozambique, migrants were settled in areas already inhabited by other ethnic groups while in South Africa the white regime predominantly settled people according to ethnicity in an attempt to create culturally homogeneous settlements commonly known as Bantustans. This chapter could have included a presentation on ethnic diversity in Mozambique, but we realized the core area of this work, the areas close to the GLTFP, does not have significant ethnic diversity.

Ethnic heterogeneity on the South African edge

In rural areas bordering the Kruger National Park (KNP), marked differences exist in ethnicity. Present-day ethnic diversity is inextricably linked to South Africa's apartheid past, which is foremost the result of government policies encouraging racial and ethnic segregation. Studying ethnicity in the Bushbuckridge area, Niehaus (2002) mentions the existence of the Tsonga, of which the Shangaan are a part, and Northern Sotho (Pulana-speaking) as the dominant ethnic groups in the area. In addition, he mentions the existence of refugees who moved into the area during the Mozambican civil war. Of significance is that Churches from Europe and America fuelled or reconfigured ethnic diversity in the area by pitting one group against another (Niehaus, 2002). At state level, ethnic reconfigurations gained momentum following the passing of the 1936 Natives Trust and Land Act, which resulted in the relocation of large groups of families (Niehaus, 2002).

Until recently, Mozambican refugees living in South Africa were not recognized as citizens despite living in South Africa for many years (Schutz, 2009). This could have had serious implications for their involvement in collective activities. However, the situation has changed as they can now access identification cards, which are crucial for accessing state grants and other forms of support.

Taking the example of Phalaborwa, Tapela *et al.* (2007) mention that communities in the area share different values and aspirations largely because of the context of it being a mining community and hence their lesser regard for wild animals.

Land restitution and the South African edge

Land restitutions in South Africa are largely based on claims by ethnically based groups of people. The claimants largely invoke common ancestral linkages, origin and language to make historical claims to place and resources located in KNP. Although Bantustans on the fringe of KNP evolved into permanent settlements based on ethnicity, changes in the South African landscape characterized by the post-apartheid era have ushered in a new dynamic in the form of land claims. Claimants to the land inside the KNP use symbols such as old burial sites, initiation sites and cattle byres to claim ownership to ancestral lands (James, 2009). The importance of the restitution exercise in South Africa for those interested in the conservation of natural resources is that there are possibilities of some ethnic groups such as the Venda in the northern part of the park demanding resettlement in the sites from which they were moved in the 1940s (Meskell, 2005).

The area around the KNP is home to several ethnic groups who continue to view certain sections of the park as their ancestral land (Meskell, 2005). Meskell cites interviews with old Venda people who still remember forced evictions and loss of property from as early as 1926 when the KNP was gazetted. Apart from the Venda and Shangaan in the northern part of the KNP, the western part of the park near Phalaborwa is home to the Pedi-speaking group who belong to the Majalati ethnic group. Ethnographic research conducted among the Majalati revealed strong and thriving connections between them and the Masorini who lived at an Iron Age site, now located inside the KNP, 800 years ago (Meskell, 2005). According to Meskell (2005), the western area is home to three land claims by different ethnic groups.

> At the time of our conversation, three separate land claims were being mounted in the Phalaborwa region by competing groups, each asserting ethnic connections to the KNP – some petitioning for a large swath across the central sector of the park, a significant portion of its present holdings. In this context

> we represented a potentially dangerous and partisan constituency, and the gathering of stories might have been viewed as a fact-finding mission from KNP, itself operating in a quasi state-like manner for much of its history.
>
> (Meskell, 2005, p. 92)

The above quote points to potential problems that may face parks in South Africa. Different ethnic groups are claiming different sites, usually using ritual to gain access to the park now and later wanting to settle permanently. To stress the point further, Meskell (2005) reports of a heritage site that was deliberately burnt at Masorini, causing dismay to both park rangers and researchers. However, despite continued claims to some areas of the KNP by communities, the South African government has ruled that no further claims will be considered. Park officials are also building bridges with neighbouring communities through community engagement programmes.

The land restitution exercise has not hidden ethnic diversity even in a small area like that in the communal land close to Pafuri Gate. Robins and van der Waal (2008) give an account of tensions between the Makuleke and Mhinga clans, with potential to undermine future viability of the Makuleke wildlife conservation project. Forced displacements of the Makuleke community from the northern part of KNP in 1969 were intended to expand the park. Today, the Makuleke are highly regarded as the model community as they have been able to reclaim the area within the park and embrace the ideals of conservation (Ramutsindela, 2002). However, Robins (2003) and Robins and van der Waal (2008) deny the existence of the homogeneous Makuleke who have been heralded in international conservation discourse, instead seeing them as an opportunistic grouping formed to access land.

Perceptions of KNP and other parks in South Africa

Just like most national parks in South Africa, the KNP is largely perceived by black South Africans who live both inside and outside as an enclave for whites and rich people (Meskell, 2005). In addition, the negative perceptions are driven by high population densities, sometimes reaching 174 people per km^2 compared with 5–10 people per km^2 on the Zimbabwean side (King and Peralvo, 2010). High population densities fuel competition for resources and often limit individual benefits from wildlife hunting programmes, hence perpetuating negative perceptions.

Ethnic heterogeneity in Zimbabwe's Southeast Lowveld

History suggests that pre-colonial warfare in Southern Africa shaped settlements of different people in the area known as the Southeast Lowveld (Mutetwa, 1976; Beach, 1984; Ranger, 1983). Following colonization, most conflicts around settlement of people lessened because of colonial subjugation and the pacification that resulted from it. Although the process of creating Native Reserves started as early as 1902, present-day settlement of different ethnic groups and people is largely a result of the movements following the enactment of the Land Apportionment Act in 1930. The Land Apportionment Act demarcated Zimbabwe, then Southern Rhodesia, into European and African areas. It was this legislation which saw large-scale displacements of Africans from their traditional homelands into Tribal Trust Lands and reserves. This exodus of people resulted in some Ndebeles being moved from Matebeleland into the Southeast Lowveld, originally settled by predominantly Shangaan, Venda and Pfumbi ethnic groups. The Karanga people, mainly from Masvingo and Mberengwa districts, followed later having been attracted into the area by good agricultural land, for both grazing and crop production. Another separate wave of Karanga comprised people who were allocated small-scale commercial farms (for the purpose of creating a black yeoman class of farmers) in the area now called Gonakudzingwa. Interestingly enough, and particularly relevant to Ndebele people, immigrants were allocated areas for which they enjoyed control through appointed kraal-heads. However, immigrants were to remain under the leadership of host tribal chiefs and headmen. This scenario is very common in areas such as Malipati and Chishinya where large groups of Ndebele people were settled after 1956 by the colonial administration. The Karanga people settled in the villages do not seem to have enjoyed the same administrative rights. They do not have autonomous villages which they control. More importantly, there has been little integration as the different groups of people have continued to pursue their independent cultures, except where intermarriage has occurred.

Natural resource utilization by different ethnic groups in the Southeast Lowveld shapes conflicts. Sola (1998) reports of conflicts between Shangaan and Ndebele ethnic groups over the harvesting and use of Ilala palm (*Hyphenea petersiana*), an important natural resource. Shangaan use the plant to produce an illicit wine, locally known as *njemani*. Ndebele people use the plant to weave baskets, mats, basins and brooms, which they sell locally and abroad. Each group of people accuses the other of using unsustainable harvesting methods that are leading to reduced regeneration of the damaged plant. Cases have also been documented of some Shangaan traditional leaders claiming ownership of all the Ilala palms, including those found in Ndebele villages and agricultural fields (Sola, 1998). The significance of these conflicts, though of low intensity, is that they have made it difficult for the Southern Alliance for Indigenous Resources (SAFIRE), a non-governmental organization, to recommend and implement strategies for sustainable management of Ilala palms (ENDA, 1993; Sola, 1998).

Cattle keeping and crop production are also shaped by ethnic diversity in areas close to the GLTFP in the Southeast Lowveld. For example, the Shangaan are widely believed to be cattle keepers. Next to keeping cattle for economic reasons, and as insurance against misfortune, cattle are important for social status. Karanga, on the other hand, are more involved in crop production for commercial purposes, growing cotton and maize. In terms of natural resources, especially wildlife, these lifestyles present challenges for TFP planners. For example, the Shangaan prefer to have more land for cattle and not wildlife. In a similar manner, the Karanga prefer more land for cropping. We must always keep in mind that these practices are ethnically constructed and well recognized in the Southeast Lowveld (Chaumba *et al.*, 2003).

Another ethnically constructed difference in the Southeast Lowveld is the way education is valued by the different people (Murungweni, 2011). For example, Shangaan people continue to have a negative appreciation of the role that education plays in promoting local development. Murungweni (2011) reports that 47 per cent of Shangaan enumerated during a survey did not attend primary school. It is common practice for young Shangaan men to leave school at primary level and cross into South Africa in search of employment. As for the girls, they are known to leave primary school soon after graduating from *komba*, a maturity ritual. Education continues to be undervalued by the Shangaan and Venda, where education levels are especially low for women. Ironically, this continues to be the case despite Shangaan informants recognizing that lack of education will lead to their children missing employment opportunities in the TFP. On the other hand, lack of education results in less appreciation of the potential economic benefits that might come with wildlife-based development projects in the area.

History and livelihoods in the Southeast Lowveld cannot be complete without reference to migration. Migration to South Africa is a well-recognized practice and has long been identified with the Shangaan and Venda (Muzondidya, 2010). In addition, migration is viewed as a maturity ritual by the Shangaan and Venda, as well as a strategy for accumulating wealth, especially cattle. Migration is a major source of remittances (cash and goods), particularly for young families.

The above reasons have serious implications for planning around natural resource management. The Shangaan have traditionally been externally oriented when it comes to meeting their economic needs and thereby likely not to perceive positive roles wildlife may play in improving welfare, as might be argued by the policy makers and TFP planners.

Ethnic heterogeneity on the edge in Dande, northern Zimbabwe

The Dande area has about 10 ethnic groups including Korekore, Karanga, Doma, Zezuru, Ndau, Ndebele, Chikunda, Zengeretsi, Chewa and Manyika. Korekore, Chikunda, Doma and the minority Zengeretsi are autochthonous while the rest are immigrants who moved into the area during the various development phases of the area, in particular during the Mid-Zambezi Resettlement Programme (MZRP) of the mid-1980s, which followed the eradication of the tsetse (Bourdillon, 1987; Derman, 1993; Hasler, 1996; Dzingirai, 1999). The Chewa originating from Malawi were formerly employed as labourers in the nearby large-scale commercial farms.

Dzingirai (1999) identified four important categories of people living in the Lower Zambezi Valley, further pointing to community heterogeneity on the edge. For example, he identified people he described as 'home seekers', and these are people largely of foreign origin who used to work in former commercial farms and moved into the area to look for homes. These people, Dzingirai (1999) argues, would probably not want wildlife in the area, a view challenging for the transfrontier conservation philosophy.

The second group of people are generally referred to as 'accumulators', largely because they are strictly not connected to the area and do not care what happens to natural resources. Accumulators were also described as opposed to CAMPFIRE and even worse, to traditional leaders because of their religious affiliation, Apostolics of the Christian Faith.

Group three comprises 'traditionalists' largely drawn from the Korekore. They prefer land-related discussions to be conducted by relevant kinship groups and more importantly, the Mhondoro cult leaders. Traditionalists support hunting by locals and have rejected CAMPFIRE for refusing to give people access to meat. This group of people has rejected fencing off wildlife areas as this is interpreted as taking away meat from people.

The fourth group of people comprise 'modernists' who are largely the young and educated farmers viewing the Mhondoro cult as counter development. The modernists are there to harness any initiative that brings financial or other returns to them and will only support wildlife initiatives if proven that they will surpass other options in revenue generation. According to Dzingirai (2008), the challenge for wildlife conservation planners would be to harmonize all these interests.

VaChikunda and Doma traditionally settled along the major rivers in Dande such as the Angwa, Kadzi, Manyame and Mwanzamutanda where they cultivated riverine plots or *dimba*. Introduction of the state-protected areas such as Mana Pools National Park and Chewore Safari Area from 1963 to 1975 pushed the early residents of these areas eastwards, towards the Angwa and further east to the Mavuradonha Escarpment (Cunliffe, 1992; Marindo-Ranganai and Zaba, 1995; Hasler, 1996). Their role as traditional hunter-gatherers was curtailed by new legislation governing hunting and foraging in wildlife protected areas. Dande ethnic homogeneity was also curtailed when the Karanga from Masvingo (were) moved into the area following the success of tsetse eradication in the mid-1980s (Derman, 1983; Mukamuri and Mavedzenge, 1997).

The Karanga group we described above was preceded by a first wave of the ethnic group into Dande between 1971 and 1973. This first group, largely from Gutu District, initially moved into the neighbouring Muzarabani (formerly Centenary) District in search of fertile land. This group was later forcibly moved to Dande by the colonial government as part of surveillance efforts to stem guerrilla incursion into the interior during the 1970s war of liberation. The second wave coincided with the MZRP. The MZRP was sponsored by the European Union and resulted in the area being attractive for cattle rearing. Migrants of the two waves of Karanga ethnic group fit well with Dzingirai's (2008) classification of modernists and entrepreneurs largely characterized by anti-conservation behaviour.

Other studies have shown that ethnicity defines access to different types of resources in the Dande edge. For example, wet and fertile riverine areas have traditionally been accessed by the Korekore and VaChikunda for production of maize, pumpkins and vegetables. Such areas are also viewed by other ethnic groups as important given the aridity of the area in general, and hence are a major source of conflict (Dzingirai, 2003).

The above-mentioned ethnic differences impact on wildlife-related conservation activities in Mbire. For example, Karanga people were allocated land in areas traditionally used by wildlife, especially elephants which also destroy crops. The Karanga have over the years engaged in large-scale cotton and maize production in traditional wildlife areas. This has led to competition over land, between people and wildlife. On the other hand, the Doma ethnic group, a minority ethnic group occupying the western part of Dande, largely remain hunter-gatherers – activities now deemed as poaching by CAMPFIRE, state and local authorities (Rural District Councils), in particular. A combination of these ethnically based activities continues to undermine CAMPFIRE activities (Mukamuri and Mavedzenge, 1997). The Karanga, traditionally agro-pastoralists, are hostile to wildlife because of the damage to crops and livestock they incur.

In Dande, conflict with the safari operators emanates from the fact that the autochthons, especially the Doma group, have for years survived as hunter-gatherers. This leads them to view safari areas and parks as infringing on their livelihood, as hunting or foraging are labelled as poaching by the officials.

Conflicts also exist around traditional beliefs which the Korekore and Doma hold and practise. Immigrants perceive such practices as strategies to dominate, subjugate and incorporate them into the Korekore and Doma religious and political system. There is contestation over ritual, political and organizational authority and legitimacy (Mukamuri and Mavedzenge, 1997; Spierenburg, 2004). The existence of such tensions and hard feelings spill into the wildlife arena because the two groups do not easily agree to common plans, and collective action is compromised. Suggestions concerning wildlife and land use by one ethnic group are often viewed with suspicion by the other.

Ethnic heterogeneity on the Hwange–Tsholotsho edge, Zimbabwe

Three dominant ethnic groups are found in Tsholotsho and Hwange areas namely San, Kalanga and Ndebele. The San are the original inhabitants of the area and their territory and community in the past extended into Botswana (Solway and Lee, 1990; Lee and Hitchcock, 2001). Tsholotsho's population breakdown is Kalanga (42.9 per cent), Ndebele (51.4 per cent) and San (5.7 per cent). The highest concentration of San is found in Ward 7, of which 6 per cent of the ward population is San. A recent survey (Le Bel *et al.*, 2011) found that Hwange Communal Area adjacent to Hwange National Park has 12 distinct ethnic groups: Ndebele, Tonga, Nambia, Chewa, Chikunda, Nyanja, Doma, Zezuru, Kalanga, Mlubale, Karanga and Shangwe with the Ndebele being the majority.

There are a more than 200 San households in Wards 7, 8 and 10 in the extreme southern and western parts of Tsholotsho District. Except in Ward 7, most of the San households in Tsholotsho are found alongside Ndebele and Kalanga households. Further, the San are on the fringes of the district's economic and socio-political systems, although they are the autochthonous people of the area. Place names such as Cawunajena, Gulalikabili, Gariya and Gibixegu denote San ancestry, with Tsholotsho District being named after a pan. Whereas the Kalanga and Ndebele are sedentary agro-pastoralists, the San were originally hunter-gatherers (Solway and Lee, 1990). The San people are traditionally dependent on natural resources, particularly wild food, and are knowledgeable about their environment and benefits they derive from it. This knowledge has earned them the reputation of being excellent wildlife trackers (Lee and Hitchcock, 2001). As their contact with the Kalanga and Ndebele increased over time, the San, facing a shortage of wild grain, increasingly bartered meat for grain such as pearl millet (*Pennisetum glaucum* (L.) R. Br) and sorghum (*Sorghum bicolor* (L.) Moench). Resource use in the area is very unbalanced in favour of the two dominant groups (Kalanga/Ndebele). San households in Tsholotsho report that they do not own any cattle. For their survival they depend on ploughing, weeding and harvesting for Kalanga and Ndebele households. They also sell thatching grass or exchange it for grain with Kalanga and Ndebele households. Occasionally, San households living close to safari concessions traded or bartered elephant meat from safari hunts for grain.

Livestock grazing constitutes a major resource use in Tsholotsho. *Lagisa* is a form of transhumance practised by livestock keepers in the communal areas. It involves the seasonal movement of cattle from one area to the other to extend the grazing range. Cattle owners or employees move into the *lagisa* area and make a temporary shelter (*umlaga*), which they abandon at the end of the grazing season. Ideally, however, the community rather than any individual owns the entire *lagisa* area. The practice is motivated by the need for reliable dry season sources of water and nutritious grazing (Madzudzo and Hawkes, 1996).

Drawing on two districts, Bulilima-Mangwe and Binga, differential benefits existed in CAMPFIRE because of planners' disregard for ethnic differences within the target communities (Madzudzo and Dzingirai, 1995). CAMPFIRE was concerned with compensating agro-pastoralists and providing them with livestock rangelands and in the process neglected peripheral San foragers. In Binga, wealthy immigrant farmers were neglected: this neglect led the affected people to develop livelihood strategies which ultimately affected CAMPFIRE negatively (Madzudzo and Dzingirai, 1995).

Similarly, the introduction of CAMPFIRE in some districts such as Tsholotsho and Bulilima Mangwe District sought to localize authority over use of wildlife, with the legal authority to manage wildlife being conferred on the Rural District Councils (RDCs) and resulted in some land use practices that were opposed to CAMPFIRE (Madzudzo and Dzingirai, 1995). The RDCs saw a source of financial and material resources in CAMPFIRE, and this gave them an incentive to control hunting by local communities effectively. In Tsholotsho, locals have been recruited and trained as game scouts to protect wildlife from poachers. With the introduction of CAMPFIRE, the political economy of wildlife conservation had taken a new turn. Through the localization of enforcement, San were completely cut off from wildlife in the name of development. CAMPFIRE has been developed in Tsholotsho with no special regard paid to the ethnic divisions of the community, and no effort has been made to target any activity specifically for the benefit of the San. This alienates and excludes traditional hunting communities such as the San, who had depended on subsistence hunting. The San were moved to Pelandaba Ward to facilitate safari hunting under CAMPFIRE. The San were thereby rendered sedentary. Lacking draught power or agricultural implements, they were left with the prospect of poor agricultural performance. In contrast with the former dispensation, in terms of which wildlife was regarded as state property, wildlife was now communal property, even though the status of the San was not equal to that of others. Couched as a community control mechanism, state surveillance had impinged on the lives of the San. Without access to wildlife or a culture of agriculture, they were reduced to beggars on the fringes of the dominant groups.

There is a high prevalence of human–wildlife conflicts in Tsholotsho. Elephants destroy crops especially along the buffer zone from the Hwange National Park. These excursions occur mainly at night. Elephants cross into the communal areas to the north and west of the district. Communities mainly affected are Gibixegu, Gulalikabili, Ndodana and Gwabazabuya. The villages along the major water sources and those bordering Hwange National Park seem to be mostly affected as elephants have been noted to travel mainly along trails winding between perennial pans in the area (Gariya, Pelandaba, Gibixegu).

Conclusion

This chapter has demonstrated that areas targeted by transfrontier initiatives are inhabited by people of different ethnic origins, a scenario likely to impinge on efforts to collectively manage natural resources, particularly wildlife. We have seen contestation and conflict over key resources such as riverine and other fertile areas. The chapter has also deliberated on ethnically driven livelihoods in different areas. For example, the Shangaan in the Southeast Lowveld in Zimbabwe are more inclined towards livestock production, while the Karanga ethnic group is more into crop production. In Mbire District, we have shown that the Doma and Zengeretsi largely depend on wildlife for their livelihoods, as opposed to Karanga and Korekore peoples who now engage in the production of cotton and maize. In the western part of Zimbabwe, San people are more dependent on wildlife than the Ndebele groups who depend more on agriculture. The San are eager to see their continued ability to hunt not wholly curtailed by authorities. The Ndebele and other small ethnic groups in the area are more interested in grazing land for their cattle, as well as land for cropping activities.

Although indigenous ethnic groups continue to dominate access to fertile riverine areas in most parts of the Zimbabwean edge, this pattern is likely to change as new social relations emerge mostly through intermarriages and the emergence of villages headed by immigrants, and in some cases, villages comprising only migrant households. Typical examples are villages and village heads in Ward 8 of Mbire District and one newly created village in Ward 11 (Masoka). However, in the interim ethnicity continues to be mobilized in resource struggles.

This chapter has also demonstrated that people living close to wildlife conservation areas do not constitute distinct cultural districts as suggested by others

(Bocchino and Murphree, 2010). Instead what we see on the ground are collages of cultures, diverging interests, livelihood trends and worldviews that differ along ethnic lines.

Ethnic configurations are partly a result of colonial manipulation rather than independent social dynamics, resulting in some ethnic groups becoming not happy with the way history has treated them and also feeling that the creation of the GLTFP poses further encroachment on their livelihoods. The groups are less likely to have positive attitudes towards wildlife conservation efforts on the edge.

We have demonstrated that ethnic differences create varying notions of authority and legitimacy over natural resource management. Autochthons largely feel that certain high-value natural resources belong to them, thereby denying access to these resources by other minority groups, especially immigrants. Education also plays a significant role in shaping local perceptions over natural resource management, particularly wildlife. The less-educated ethnic group members are less likely to get well-paid (formal) employment. As a result, they may be forced by diminished opportunities to engage in exploiting wildlife for sustenance, including poaching. Migration to neighbouring countries by certain ethnic groups in search of employment, especially by the youth, creates a less appreciative attitude towards natural resources, and wildlife in particular.

In addition, ethnic differences have played out in local leadership structures by weakening traditional institutions, which could otherwise form the cornerstone of locally driven natural resources management practices. A weakening of traditional leadership structures undermines the cardinal pillars of collective action, including organization, networking, mobilization and engaging in joint activities. We therefore strongly argue for a natural resources development framework that recognizes and plans around ethnic diversity at the edge. We also recognize that it will be difficult for policy makers working at national levels to develop policies that target specific ethnic groups. However, the way forward needs to start with understanding various ethnic groups at the edge, followed by dialogue and negotiation, both key in bringing about compromises. Understanding cultural diversity helps to uncover the underlying causes of conflicts that result in increased vulnerability of poorer households to forces that play out on the edge.

References

Alexander, J. and McGregor, J. (2000) 'Wildlife and politics: CAMPFIRE in Zimbabwe', *Development and Change*, vol. 31, no 3, pp. 605–627

Balint, P.J. and Mashinya, J. (2008) 'CAMPFIRE through the lens of the "commons" literature: Nyaminyami Rural District in post-2000 Zimbabwe', *Journal of Southern African Studies*, vol. 34, no. 1, pp. 127–143

Barth, F (1966) *Ethnic Groups and Boundaries: The Social Organisation of Culture Difference*, Universitetsforlaget, Waveland Press, Oslo

Beach, D.N. (1984) *Zimbabwe Before 1900*, Mambo Press, Gweru

Bocchino, C. and Murphree, M. (2010) 'Overlapping cultural commons and districts in the Great Limpopo Transfrontier Conservation Area: potentials for local economic development', Working Paper 12/2010, CSS, Universita' di Torino, Turin

Bourdillon, M.F.C. (1987) *The Shona Peoples: An Ethnography of the Contemporary Shona, with Special Reference to their Religion*, Mambo Press, Gweru

Chandra, K. (2011) 'What is an ethnic party', *Party Politics*, vol. 17, no. 2, pp. 151–169

Chaumba, J., Scoones, I. and Wolmer, W. (2003) 'New politics, new livelihoods: changes in the Zimbabwean lowveld since the farm invasions of 2000', *Review of African Political Economy*, vol. 98, pp. 585–560

Connor, T.K. (2003) 'Crooks, commuters and chiefs, home and belonging in a border zone in Pafuri, Gaza Province, Mozambique', *Journal of Contemporary African Studies*, vol. 21, no. 1, pp. 93–120

Cunliffe, R. (1992) 'An ecological resource survey of the communal lands of Centenary District', WWF Multi-species Paper 26, Harare

Derman, B. (1993) 'The unsettling of the Zambezi Valley: an examination of the Mid-Zambezi Rural Development Project', *CASS Occasional Papers*, Centre for Applied Social Sciences, University of Zimbabwe, Harare

Dzingirai, V. (1996) '"Every man must resettle where he wants": the politics of settlement in the context of community wildlife management programme in Binga, Zimbabwe', *Zambezia*, vol. 23, no. 1, pp. 19–30

Dzingirai, V. (1999) '"This good land is not for elephants": poverty, migration and development in the Zambezi Valley, Zimbabwe', *Journal of Social Sciences*, vol. 3, no. 4, pp. 265–271

Dzingirai, V. (2003) '"CAMPFIRE is not for Ndebele migrants": the impact of excluding outsiders from CAMPFIRE in the Zambezi Valley, Zimbabwe', *Journal of Southern African Studies*, vol. 29, no. 2, pp. 445–459

Dzingirai, V. (2008) 'The outlet valve: the role of kinship in facilitating mobility among migrants from Zimbabwe', *Zambezia*, vol. 35, no. 1/2, pp. 21–39

ENDA (1993) 'Socio-economic and natural resource survey of Sengwe communal lands, Zimbabwe', Government of Zimbabwe, Harare

Ford, C.L. and Harawa, N.T. (2010) 'A new conceptualisation of ethnicity for social epidemiologic and health equity research', *Social Science & Medicine*, vol. 71, pp. 251–258

Hasler, R. (1996) *Agriculture, Foraging and Wildlife Resource Use in Africa: Cultural and Political Dynamics in the Zambezi Valley*, Kegan Paul International, London

Hazarika, S. (2004) 'Land, conflict, identity in India's north-east: negotiating the future', *Futures*, vol. 36, pp. 771–780

James, D. (2009) 'Burial sites, informal rights and lost kingdoms: contesting land claims in Mpumalanga, South Africa', *Africa*, vol. 7, no. 2, pp. 228–251

King, B. and Peralvo, M. (2010) 'Coupling community heterogeneity and perceptions of conservation in rural South Africa', *Human Ecology*, no. 38, pp. 265–281

Le Bel, S., Murwira, A., Mukamuri B., Czudek, R., Taylor, R. and La Grange, M. (2011) 'Human wildlife conflicts in Southern Africa: riding the whirlwind in Mozambique and Zimbabwe', in J.L. Pujol (ed.), *The Importance of Biological Interactions in the Study of Biodiversity*, vol. 15, pp. 284–322

Lee, R.B. and Hitchcock, R.K. (2001) 'African hunter-gatherer: survival, history and the politics of identity', *African Study Monographs, Supplement*, vol. 26, pp. 257–280

Lee, Y. (2009) 'Migration, migrants, and contested ethno-nationalism in Korea', *Critical Asian Studies*, vol. 41, no. 3, pp. 363–380

Logan, B.I. and Moseley, W.G. (2002) 'The political ecology of poverty alleviation in Zimbabwe's Communal Areas Management Programme for Indigenous Resources (CAMPFIRE)', *Geoforum*, vol. 33, no. 1, pp. 1–14

Madzudzo, E. and Dzingirai, V. (1995) 'A comparative study of the implications of ethnicity on CAMPFIRE in Bulilima Mangwe and Binga districts of Zimbabwe', *Zambezia*, vol. 22, no. 1, pp. 25–42

Madzudzo, E. and Hawkes, R. (1996) 'Grazing and cattle as challenges in community based natural resources management in Bulilimamangwe District of Zimbabwe', *Zambezia*, vol. 23, no. 1, pp. 1–18

Marindo-Ranganai R. and Zaba, B. (1995) 'Animal conservation and human survival: a case study of the Tembomvura people of Chapato Ward in the Zambezi Valley – Zimbabwe', Centre for Applied Social Sciences, University of Zimbabwe, Harare

Marshall, G. (1998) 'Ethnicity', in *A Dictionary of Sociology*. Available online: www.encyclopedia.com/doc/1O88-ethnicity.html (accessed 16 November 2011)

Meskell, L. (2005) 'Archeological ethnography: conservations around the Kruger National Park', *Archeologies*, vol. 1, no. 1, pp. 81–100

Mukamuri, B.B. (1989) 'Rural conservation strategies in south central Zimbabwe: an attempt to describe Karanga thought patterns, perceptions and environmental control', in R. Grove (ed.), *Conservation and Rural People in Zimbabwe*, Cambridge African Monographs and Baobab Press, Cambridge, UK, pp. 139–149

Mukamuri, B.B. and Mavedzenge, T. (1997) 'People and big animals: an evaluation of the CAMPFIRE project in Guruve district', CIRAD, Harare

Mukamuri, B.B., Campbell, B.M. and Kowero, G. (2000) 'Local organisations and natural resource management in the face of economic hardships: a case study from Zimbabwe', in G. Kowero, B.M. Campbell and R. Sumaila (eds), *Policies and Governance Structures in Woodlands of Southern Africa*, Centre for International Forestry Research, Bogor, Indonesia, pp. 16–44

Murombedzi, J. (1992) 'Decentralization or recentralization? Implementing CAMPFIRE in the Omay Communal Lands of the Nyaminyami District', CASS Working Paper, Center for Applied Social Sciences, University of Zimbabwe, Harare

Murombedzi, J. (1999) 'Devolution and stewardship in Zimbabwe's CAMPFIRE programme', *Journal of International Development*, vol. 11, no. 2, pp. 287–293

Murphree, M. (2001) 'Community, council and client: a case study in ecotourism development from Mahenye, Zimbabwe', in D. Hulme and M. Murphree (eds), *African Wildlife and Livelihoods: The Promise and Performance of Community Conservation*, James Currey, Oxford, UK, pp. 177–194

Murungweni, C. (2011) 'Vulnerability and resilience of competing land-based livelihoods in south eastern Zimbabwe', Ph.D. thesis, Wageningen University, Wageningen

Musumali, M.M., Larsen, T.S. and Kaltenborn, B.P. (2007) 'An impasse in community based natural resource management implementation: the case of Zambia and Botswana', *Oryx*, vol. 41, no. 3, pp. 306–313

Mutetwa, R.M.G. (1976) 'The "political" and economic history of the Duma people of south-eastern Rhodesia from the early eighteenth century to 1945', Ph.D. thesis, University of Rhodesia, Harare

Mutimukuru-Maravanyika, T. (2010) 'Can we learn our way to sustainable management? Adaptive collaborative management in Mafungautsi state forest, Zimbabwe', Ph.D. thesis, Wageningen University, Wageningen

Muzondidya, J. (2010) 'Makwerekwere: migration, citizenship and cultural identity among Zimbabweans in South Africa', in J. McGregor and R. Primorac (eds), *Zimbabwe's New Diaspora: Displacement and the Cultural Politics of Survival*, Berghahn Books, Oxford, UK, pp. 37–58

Niehaus, I. (2002) 'Ethnicity and the boundaries of belonging: reconfiguring Shangaan identity in the South African lowveld', *African Affairs*, vol. 101, no. 405, pp. 557–583

Ostrom, E. (2007) 'A diagnostic approach for going beyond panaceas', *Science*, vol. 104, no. 39, pp. 15181–15187

Ostrom, E. (2009) 'A general framework for analysing sustainability of socio-ecological systems', *Science*, vol. 325, no. 5939, pp. 419–422

O'Sullivan, J. and Dana, T. (2008) 'Redefining Maori economic development', *International Journal of Social Economics*, vol. 35, no. 5, pp. 364–379

Phimister, I. (1988) *An Economic and Social History of Zimbabwe, 1890–1948: Capital Accumulation and Class Struggle*, Longman, London

Powell, J. (2000) 'Expanding horizons: environmental and cultural values within natural boundaries', *International Journal of Heritage Studies*, vol. 6, no. 1, pp. 49–65

Ramutsindela, M. (2002) 'The perfect way to ending a painful past? Makuleke land deal in South Africa', *Geoforum*, vol. 33, pp. 15–24

Ranger, T. (1983) 'The invention of tradition in colonial Africa', in E.J. Hobsbawn and T. Ranger (eds), *The Invention of Tradition*, Cambridge University Press, Cambridge

Ranger, T. (1985a) *Peasant Consciousness and Guerilla War in Zimbabwe: A Comparative Study*, James Currey, London

Ranger, T. (1985b) 'The invention of tribalism in Zimbabwe', Mambo Occasional Papers, issue 19. Mambo Press, Gweru

Ranger, T. (1999) *Voices from The Rocks: Nature, Culture and History in the Matopos Hills of Zimbabwe*, James Currey, Oxford

Robins, S. (2003) 'Whose modernity? Indigenous modernities and land claims after apartheid', *Development and Change*, vol. 34, no. 2, pp. 265–285

Robins, S. and van der Waal, K. (2008) 'Model tribes and iconic conservationists? The Makuleke Restitution Case in Kruger National Park', *Development and Change*, vol. 39, pp. 53–72

Schutz, E.J. (2009) 'Reframing vulnerability: Mozambican "refugees'" access to state-funded pensions in rural South Africa' *Journal of Cross Cultural Gerontology*, vol. 24, pp. 241–258

Sola, P. (1998) 'The performance of *Hyphaene petersiana* under community-based utilisation in Sengwe communal lands, Zimbabwe', IES Working Paper, University of Zimbabwe, Harare

Solway, J.S. and Lee, R.B. (1990) 'Foragers, genuine or spurious? Situating the Kalahari San in history', *Current Anthropology*, vol. 31, no. 2, pp. 109–146

Songorwa, A.N. (1999) 'Community-based wildlife management (CWM) in Tanzania: are the communities interested?', *World Development*, vol. 27, no. 12, pp. 2061–2079

Spierenburg, M.J. (2004) *Strangers, Spirits, and Land Reforms: Conflicts about Land in Dande, Northern Zimbabwe*, Brill, Leiden

Tapela, B.N., Maluleke, L. and Mavhunga, C. (2007) 'New architecture, old agendas: perspectives on social research in rural communities neighbouring the Kruger National Park', *Conservation and Society*, vol. 5, no. 1, pp. 60–87

Virtanen, P. (2005) 'Tradition, custom, and otherness: the politics of identity in Mozambique', *Identities*, vol. 12, no. 2, pp. 223–248

Wallace, M. (2003) 'Making tradition: healing, history and ethnic identity among Otjiherero-speakers in Namibia, c. 1850–1950', *Journal of Southern African Studies*, vol. 29, no. 2, pp. 355–372

Werbner, R. (1995) 'In memory: a heritage of war in south-western Zimbabwe', in N. Bhebe and T. Ranger (eds), *Society in Zimbabwe's Liberation War, Volume II*, University of Zimbabwe Publications, Harare, pp. 192–205

6 On the edge of state and economy

Vupenyu Dzingirai, Jens A. Andersson, Frédéric Baudron, Jessica Milgroom, Chrispen Murungweni and Xavier Poshiwa

Introduction

The previous chapters discussed the constitution of 'the edge' (Chapter 3) and the diversity of people's activities (Chapters 4 and 5) in these areas in relation to its major defining characteristic: its proximity to wildlife conservation areas. This chapter shifts the attention to another defining characteristic: the position of 'the edge' in border zones, at the margins of the state and national economy. As has already been elaborated, protected wildlife areas in Southern Africa were often declared in dry and disease-ridden areas where settlers saw no mining prospects or potential for commercial agriculture. Not surprisingly then, these uncontested and sparsely populated areas became the border zones of the new states that the European powers defined.

From being regarded as marginal and even dangerous zones threatening the political economy of the state's core in early colonial times, 'the edge' has increasingly become redefined as an area of opportunity. In this chapter, we focus on this apparent paradox of marginality and opportunity, looking mostly at the Zimbabwean edge. It is shown how different social actors redefined the edge as areas of opportunity, thus strengthening the ambiguity of the edge. For instance, local peoples, migrant farmers and other entrepreneurial-spirited came to value the edge for its potential for natural resource harvesting, expanded agricultural production or the opportunities the crossing of international boundaries offer. State and non-governmental agencies, on the other hand, saw opportunities for development. By contrast, for the politically or economically marginalized, the edge may not so much have represented opportunity, but was valued nevertheless as a hiding place.

First, we analyse how different valuations and perceived opportunities during the early colonial period have been translated into legal and extra-legal activity on the edge. We then focus on the contradictions of (state) interventions and policy at the edge and how these perpetuate its ambiguity. The chapter concludes that the persistent ambiguity of the edge, shaped by its particular embeddedness in wider developments, is likely to deflect planned trajectories of mainstreaming and incorporation, including TFCA proposals. Frequently imagined to yield social and ecological benefits on a large scale, these TFCAs are likely to be re-routed to something they were not intended to be.

Valued marginality

The edge as buffer zone

As already elaborated in Chapter 3, disease threats for humans and livestock, harsh climatic conditions and sometimes poor soils were often major factors in defining what were to become protected wildlife areas. Preoccupied with developing settler agriculture in areas with no mining prospects, both settlers and their governments saw these dry and disease-ridden territories as marginal farming areas and had little interest in developing them. While in South Africa a number of protected wildlife areas were established in the 1920s and 1930s with the aim of enclosing endangered species (Carruthers, 2009, p. 38), in Rhodesia the situation was different. Up to as late as the 1950s these areas were often perceived as a threat to the development of the economy. Seen as reservoirs of tsetse fly, areas such as the Zambezi Valley and parts of the Limpopo River basin posed a threat for the emerging European livestock industry (Phimister, 1978). In Rhodesia the settler state therefore attempted to control such disease threats by creating wildlife-free corridors using game-proof fences, such as the one parallel to the Zambezi Valley escarpment. By systematically eliminating wild herbivores in these corridors, it sought to contain the spread of tsetse fly from infested wildlife areas (Child and Riney, 1987; see also Chapter 3). In other areas, such as along the Zambezi River, the Tonga people were forced to move out (Dzingirai, 2003). Of little productive value to the expanding colonial economy, the areas at the margins of the newly formed colonial states were thus 'split' into areas mostly valued by European settlers as recreational hunting grounds, and their fringes bordering African areas and settler farms. The latter were merely valued for their role as buffer zones that protected an emerging agricultural sector. Perhaps ironically, native people valued these margins of the state and the economy for rather different reasons.

The edge as hiding place

Before the advent of colonialism, the areas that were to become border zones of colonial states were already valued as hiding places. For instance, in the 1880s, Tsonga- or Shangani-speaking groups fled the Luzo-Gaza wars further south, and settled in the Zimbabwean lowveld (Bannerman, 1978; Niehaus, 2002). Before this, slave trading on Africa's east coast had also forced people into such marginal areas where they could not be easily detected by Arab ivory and slave hunters, Portuguese slave dealers and collaborating local strongmen (Niehaus, 2002).

With European colonizers extending their control over territory and existing African chiefdoms, the edge of newly formed states came into existence.

Such colonial state formation reinforced these areas' role as hiding places for those marginalized by it. An early example is King Lobengula, the last Ndebele leader who led the resistance to European occupation in the Matabele war of 1893. As his kingdom was overthrown, he eluded government soldiers by fleeing into the woodlands north of the Shangani River (Alexander *et al.*, 2000).

Much later, in the 1970s, such areas at the margins of the state again became significant as hiding places for those resisting colonial rule. For instance, guerrillas infiltrating the Rhodesian heartland from Mozambique, Zambia and Botswana concealed themselves here. The Mid-Zambezi Valley is a case in point. It became the first operational zone after the liberation war intensified in 1972 (Martin and Johnson, 1981; Mtisi *et al.*, 2009). The guerrilla army executed swift attacks on farms on the plateau, returning to the forests of the valley for cover. While the guerrillas mobilized local people for their war effort and built on their knowledge of the area (Lan, 1985), the Rhodesian state used the area to experiment with new methods of guerrilla suppression. For instance, 'a broad strip of land along the border was declared a no-go area and defoliated' (Weinrich, 1977, p. 210). People's cattle were sold off, their villages destroyed and they were forcefully resettled in so-called 'protected villages', to prevent them from supporting the unprovisioned guerrillas. The remoteness, high degrees of poverty and the general low level of education in the Zambezi Valley facilitated the enforcement of such drastic measures. In such remote border zones, the colonial state could use forced labour and more stringent curfew measures than in more central areas, as news on such measures could more easily be suppressed (Weinrich, 1977). In the Southeast Lowveld, where protected villages were established in the mid-1970s, anti-personnel mines were used to prevent encroachment by the guerrillas. Valued as hiding places, the very remoteness of these border zones also enabled state agencies to implement their more covert operations. As the insurgency spread, it largely withdrew from these areas, deploying a strategy of swift aerial raids.

The edge as a hiding place in post-colonial Zimbabwe

After Zimbabwe's independence, the edge continued to provide shelter to those on the run from the state, as evidenced by the insurgency of political opponents and the subsequent terror campaign of the national army's Fifth Brigade in Matabeleland in the period 1983–1986. Earlier clashes between government forces and elements of the former liberation army that had operated in the western part of the country, as well as apartheid South Africa's destabilization policies, were used as justification for army intervention (Phimister, 2008). But the army's extremely violent '*Gukurahundi*' campaign assumed ethnic and regionalist dimensions as well, and left an estimated 20,000 people dead (CCJP, 1997; Muzondidya, 2009, p. 184). Insurgents planning attacks, as well as rural villagers attempting to flee the violence, took refuge in the unoccupied hills and woodland areas of Matabeleland. The edge's role as the locale of the politically and economically marginalized was thus reinforced, while the state's dealings with this 'Matabeleland crisis' strengthened the ethnic dimension of this marginalization.

In recent years the edge has again provided refuge to the politically and economically marginalized. This time it was citizens fleeing from militias in defence of the collapsing Zimbabwean state. Following government-sponsored 'land invasions' and subsequent land reform in the 2000s, many farm workers were displaced from white-owned commercial farms (Marongwe, 2007). Either believed to be sympathetic to their white employers, or simply branded as 'foreigners' as many are descendants from immigrants and have no rural homes within Zimbabwe to return to, they were forced to resettle. Many descended into the Zambezi Valley, negotiating for land with traditional leadership and youths who, because of state connections, now wielded power as 'war veterans' (Nemarundwe, 2005). Some settled within existing villages, but others established new settlements in areas closer to the boundaries of parks and safari areas, interfering with conservation initiatives organised around long-term residents (cf. Dzingirai, 2003).

The edge as labour reserve

As all rural areas occupied by Africans, or designated for African occupation by segregationist policies, the edge has been valued as a labour reserve. Forced or voluntary, labour recruitment for emergent capitalist industries has been a salient feature of these areas, be it for gold mining around Johannesburg, railway development, or large-scale farms, mines and urban centres that developed elsewhere in the Southern African region (Wolpe, 1972; Murray, 1995). All such centres of employment looked to edge areas as a source of cheap labour.

International labour migration was periodically regulated by inter-state agreements and agreements that included major recruiting agencies, such as the recruitment arm of the South African mining industry, the Witwatersrand Native Labour Association (WNLA) established in 1901. These agreements stipulated among others recruitment quota, deferred pay to workers, and payments to governments of labour-exporting countries (Paton, 1995). Colonial governments of labour-exporting countries such as Mozambique were thus paid per recruited migrant worker by the South African gold mining industry (Harries, 1994; Lundstrum, 2007). Colonial governments tried to control migratory flows and create revenues from such movements by instituting pass systems and providing 'free' transport, such as the 'Ulere' system in Southern Rhodesia, which included boat ferries in the Mid-Zambezi Valley:

> The Ulere service was first established in 1925 . . . for the purpose of assisting voluntary (non-recruited) labour on its way to the Colony to work. [It] commenced with a few ferries on the middle Zambezi . . . During 1951 Ulere handled 61% of male migrants who arrived in the Colony. The cost of the service is incredibly low when compared with the number of extra labour days the Colony thereby gains.
>
> (Annual report of the Commissioner of Native Labour, 1951, NAZ, file S1051)

Migrants often preferred to avoid or escape such official channels in search of higher wages, more freedom of movement and to evade the colonial tax order (Andersson, 2002, 2006; Newitt, 1995). They would cross the loosely controlled borders illegally, risking capture by unscrupulous labour recruiters in search of 'capitation' fees from the labour demanding industries (Murray, 1995), and state agencies trying 'to control such illegal activities', but meanwhile 'redirecting' migrants into the official channels leading to centres of high labour demand. Hence, border zones became transit areas and attempts to regulate migratory flows formed a major incentive for the state to control the border area with an increased presence (Lundstrum, 2007).

The edge as safety valve

While various people may have valued the margins of the state and national economy as a hiding place, the counterpoint to this is that the state and powerful elites have often regarded the edge as a 'dumping ground' for what they considered 'excess peoples'. The most poignant example of this valuation of the edge as safety valve is South Africa's 1913 Land Act, the Group Area Act of 1950 and the apartheid governments' subsequent Bantustan policies. It resulted in over 3.5 million people being forcibly removed to their designated 'Group Area', often without having any link to these areas other than an ascribed ethnicity (Fabricius and de Wet, 2002, p. 142). Thus the South African state created discontinuous 'self-governing homelands' and also removed people from conservation areas such as the Kruger National Park into homelands such as Venda, Gazankulu, Lebowa and Kangwane situated on the edge of it. It aggravated unsustainable resource use in these densely settled areas, and instilled negative attitudes towards conservation among those forcibly removed (Fabricius and de Wet, 2002).

A similar displacement of people took place in colonial Zimbabwe, albeit without attempts to form ethnically homogeneous and self-governing areas. The Rhodesian state mainly sought to off-load excess population from prime agricultural land to give way to commercial farms and ranches, but the numbers of people involved were far less than in South Africa. Karanga-speaking peoples from Masvingo moved to the sparsely populated Zambezi River Valley, such as the Gokwe District and the Mid-Zambezi Valley in the 1960s. Such migrations, voluntary or state-enforced, were often instigated by a lack of opportunity in the densely populated sending areas, and often gave rise to chain migrations as newcomers joined family members who had migrated earlier (Nyambara, 2001; see also Chapter 4). Even though conflicts between autochtons and immigrants would often arise at some stage, in most areas, migrants were initially welcomed by long-term residents as they would help clear the bush and in the process drive away wild animals (Spierenburg, 1995). Migrants also increased the following of tribal authorities (Dzingirai, 2003).

As anti-colonial protest increased, sparsely populated edge areas not only became more important as a hiding place for the insurgency (see above), but they were also used to isolate nationalist leaders considered a threat to state security.

For instance, Joshua Nkomo and other nationalist leaders were kept at restrictive camps such as Gonakudzingwa in the Southeast Lowveld (Nkomo, 1984).

Marginality and opportunity

An 'empty space' appropriate for state-controlled development

Regarded a threat to the settler economy, and acting as a safety valve, the settler state also valued the 'emptiness' of edge. This is exemplified by early maps of areas such as the Zambezi Valley and the Southeast Lowveld, which show large blank or 'unassigned' areas (Floyd, 1962; Palmer, 1977, see Figure 3a). The long-term presence of people in these areas – though their number was relatively small – was simply ignored. For the colonial authorities, the issue was merely whether the 'unassigned' areas would be able to carry a substantial population in the future.

Defined as 'unassigned', 'empty' and 'marginal',[1] edge areas seemed open for new initiatives, which simultaneously meant increased state control over these areas. State farms and small-scale irrigation schemes for Africans constituted the first of such initiatives in Southern Rhodesia. Although records are vague, it appears the first planned schemes in the Mid-Zambezi Valley date back to the 1930s (Biodiversity Project, 2002). From the early 1940s, government also developed irrigation schemes in the Save Valley in south-eastern Zimbabwe (Roder, 1965).

Later, in the 1960s, the state stimulated settlement in the vast sparsely populated areas along Lake Kariba and the Zambezi River valley. Tsetse eradication, the provision of transport for migrants and the provision of boreholes in areas such as the present-day Gokwe District constitute another way in which the edge of the state became the focus of state-directed government.

A third state initiative that affected the edge was the Native Purchase Areas that aimed to create a class of African yeoman farmers. Established from the 1930s onwards, Gonakudzingwa Native Purchase Area bordering (what is now) the Gonarezhou National Park, is an example of this policy dating back to the 1960s (Wright, 1972).

To be sure, state-led development initiatives on the edge, ranging from borehole drilling to Purchase Areas and irrigation schemes, was usually not only motivated by a concern for African agricultural development. Initiatives to intensify land use simultaneously served its segregationist policies. By increasing the carrying capacity of the land, the state could settle more people in such designated African areas (Andersson, 2002). Thus 'development projects' often conflated different values attributed to the edge – the edge as 'safety valve' and 'area for development'.

As in colonial Zimbabwe, in Portuguese East Africa, the edge was seen by the state as an appropriate zone for state-controlled agricultural schemes. An example is the state's development of an irrigation scheme at Chokwe in the Limpopo River basin (Lundstrum, 2007) and the construction of a dam on the Rio Elephantes (now the southern boundary of the Limpopo National Park) at Massingir to provide the water for it. Work on the dam began in 1970, but was only completed after independence.

Not only the colonial state regarded the edge an empty space representing an opportunity for development. White settlers and their descendants did as well. For instance, they established sugar estates in the Southeast Lowveld after World War II (de Jager, 1988). This initiative triggered numerous state-supported infrastructural developments (Saunders, 1989), such as the building of the Lake Kyle dam (now Muturikwi) near Masvingo that was to supply the water to the sugar estates. Other dams were to follow, as was road and railway infrastructure. The sugar estates created employment for local people as well as migrant workers, and resulted in the development of Chiredzi, Triangle and Hippo Valley towns.

In northern South Africa, areas designated for African occupation generally did not extend to the country's borders. This was a deliberate policy, resulting in the state taking over the area near Pafuri and creating a military zone along the Limpopo River, the Madimbo Corridor. The area between Beitbridge and Pafuri was thus placed under direct state control or occupied by commercial farm enterprises. This chequered landscape of commercial farms and military zones extended also to the border area to the west of Beitbridge. Thus the apartheid state sought to control illegal movements across the border, while commercial farms in this border zone continue to tap a source of cheap labour: illegal border crossers.

The edge and the post-independence state: agricultural development

The double role of the edge as the state's safety valve and 'empty' areas for development continued after independence. The areas at the margins of the state and national economy remained the focus of state projects, as evidenced by state attempts to encourage the land hungry to settle into edge areas such as the Zambezi Valley (Derman, 1993).

In independent Mozambique, the ruling FRELIMO party devised a strategy of 'socialization of the countryside' in the 1970s. This meant a radical restructuring of the rural economy, including that of border zones (Lundstrum, 2007). Dispersed rural households were concentrated into villages to facilitate the provision of services and to work on large-scale, mechanized state farms that often replaced equally ambitious Portuguese plantations. In the area that was to become the edge of Limpopo National Park, people were not only confronted with forced villagization but, a bit later, also with a civil war, forcing them to flee to urban centres, or across the border to South Africa. As a result of these developments, the sparsely populated area adjacent to South Africa's Kruger National Park was further depopulated, only to be repopulated after Mozambique's civil war ended in 1992. Designated as a hunting concession (Coutada 16) in the colonial period, a large part of this area is characterized by water scarcity and rocky uplands, causing human settlement to concentrate along the Limpopo, Shingwedzi and Elephants rivers. This enabled international conservationists to portray the area as largely untouched, despite the presence of 27,000 people. As Coutada 16 has a rich biodiversity (Wolmer, 2003; Duffy and Slotow, 2006), the Mozambican government was convinced by its international partners in conservation, including the South African government and the Peace Parks Foundation, to declare the hunting concession a national park in the late 1990s, forming the third leg of the tri-national GLTFP. Almost a decade later, while facing the challenge of increased human–wildlife conflict along the Elephants and Limpopo Rivers where some 20,000 people live, the Limpopo National Park remains a largely underdeveloped state dominated project, which is sustained only through international donors.

In contrast to Mozambique, the post-colonial state in Zimbabwe envisaged the edge as a zone open to agricultural development, thus continuing older policies that sought to increase the carrying capacity of these agriculturally marginal areas that constituted the edge. For instance, the Zambezi Valley became a target for massive resettlement. Underdeveloped and with relatively small populations, these areas were viewed as capable of absorbing excess populations from the densely populated Communal Areas elsewhere in the country (Spierenburg, 1995). The Mid-Zambezi Rural Development Project (MZRDP), funded by the African Development Bank and the government of Zimbabwe, sought to develop the Mid-Zambezi Valley and resettle 3,000 families, in addition to an estimated 4,600 'frontier migrants' – people who had come to the Valley after restrictions on mobility were lifted at independence (Aubin, 1997; Ivy, 1998). They continued a process of chain migration to the edge that had started decades earlier.

State planning of the resettlement process was detailed: not only were residential plots, arable lands and garden areas pegged, resettled families' livestock holdings were restricted as well, and subdivisions of the allocated farms prohibited (RTTCP, 1995; Ivy, 1998). But as elsewhere where the state tried to impose larger control over the edge, those resettled subverted the project's regulations, particularly in relation to livestock holdings (RTTCP, 1995; Aubin, 1997). They saw the opportunities of the edge differently.

Opportunistic appropriation from below

This section turns to the ways in which local peoples have defined the edge as a place of opportunity and wealth accumulation, redefining the state plans for such areas. These opportunities were both legal and extra-legal, and are described separately in the sections below.

Cotton production

While land was abundant in the Mid-Zambezi Valley, wartime mobility restrictions, human disease, tsetse-fly-transmitted sleeping sickness (bovine trypanosomiasis), as well as low and erratic rainfall had limited population movements into it. But after tsetse eradication campaigns commenced in some parts in 1983, more people began to see the Valley as an opportunity for wealth accumulation (Pollock, 1991; Aubin, 1997). Cotton, a crop introduced in the 1960s on state farms and extended to smallholder farmers in the Gokwe District (Reid, 1976;

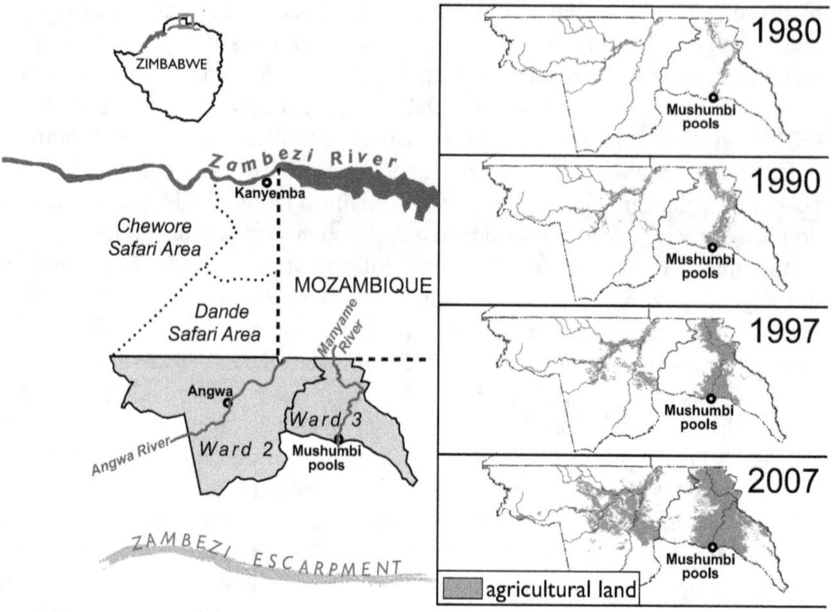

Figure 6.1 Expansion of agricultural land (fields and fallows) into the interfluves of the Mid-Zambezi Valley, 1980–2007.

Worby, 2000), became the vehicle of wealth accumulation. Migration into the Mid-Zambezi Valley almost doubled the population of Dande Communal Land from 36,074 to 71,096 between 1992 and 2002. Population increased faster in eastern wards targeted by operations of tsetse control (137 per cent in Ward 3) than in the more western wards, not targeted by these projects (80 per cent in Ward 2). In the 2006/7 season a survey across Wards 2, 3 and 9 revealed that 64 per cent of the 176 interviewed heads of household were migrants, originating from outside Dande Communal Land (Figure 6.1) (Baudron *et al.*, 2011). Another wave of migrants occurred after the 2000 land reform, represented mainly by former farm workers from large-scale commercial farms of the nearby plateau.

A similar process is discernible in the upper Zambezi Valley, the vast area along the shores of Lake Kariba. In the late 1980s, regional politicians encouraged the settlement of this vast and sparsely populated land in which there was no European settlement (Cumming and Lynam, 1997). Through state media, migrants were lured from Masvingo Province, and the Gokwe and Lupane districts. Ethnic integration was sometimes used by settlement promoting politicians to silence imminent protest by the indigenous Tonga peoples of these areas (Dzingirai, 1998).

Hence, as in Gokwe District during the colonial period, after independence both the state and migrants saw the edge as a place of opportunity. Cotton production drove settlement and agricultural development on the edge, and the state supported it through agricultural extension services and the provision of credit through the Agriculture Finance Corporation (Aubin, 1997).

Commodity smuggling

Not all initiative from below was legal. Border zones are, by their very nature, prone to cross-border smuggling and other illegal activity. As already indicated above, for migrant workers in search of better wages across national borders, the border zone presented a risk of captivation by labour touts soon after the establishment of national borders. Labour touts could often get away with their illegal practices, crossing the national border to avoid arrest (Murray, 1995). In addition, state presence in the form of permanent police or administrative posts was usually limited in these areas. For instance, in the first decade of the twentieth century, South Africa's nearest police base was three days' ride from Pafuri, the area where Rhodesia, South African and Portuguese East Africa meet (Jeeves, 1985).

Besides labour, illegal cross-border trade in commodities has always been an important opportunity on the edge – for wealth accumulation or sheer survival. A most poignant example is the upsurge in illegal cross-border trade on Zimbabwe's borders, after the country entered a deep economic crisis in 2000. Shortages within Zimbabwe resulted in illegal grocery imports from Zambia through Kanyemba on the Zambezi River, and movements of cattle and grain into Zambia, where these could be sold in foreign currency. The same practice took place in the Southeast Lowveld of the country where households smuggled basic commodities such as sugar and flour from Mozambique and South Africa. In addition, local peoples on either side of the borders linked up and organized illegal 'border-jumping' and onward transport in South Africa, thus intensifying the smuggling of people from Zimbabwe. Through their intricate knowledge of the borders, edge residents carefully steered the clients away from dangerous routes where they would encounter security agents.

Cattle theft and cross-border trade

As with illegal cross-border migration and commodity trade, cattle theft and rustling is not a new phenomenon on the edge, as Figure 6.2 shows. It has been part and parcel of the livestock economy, and increased border controls and fences have not been able to eliminate it. At the other extreme, a highly regulated formal marketing system, with the state-controlled Cold Storage Commission (CSC) as the predominant buyer, has played an important role in suppressing cattle theft and rustling. For instance, the introduction of the Economic Structural Adjustment Programme of 1991, which opened up the cattle marketing system to a larger range of players, did not seem to have a major effect on cattle thefts in the Southeast Lowveld. Although there were major shifts in production and marketing, such as a rapidly declining market share of the CSC from about 80 per cent in 1990 to less than 40 per cent by 1995, a growing share of Communal Area cattle in total cattle holdings, and a continued decline of the large-scale commercial cattle sector (Mavedzenge *et al.*, 2008, p. 617), reports of cattle theft did not increase (Figure 6.2).

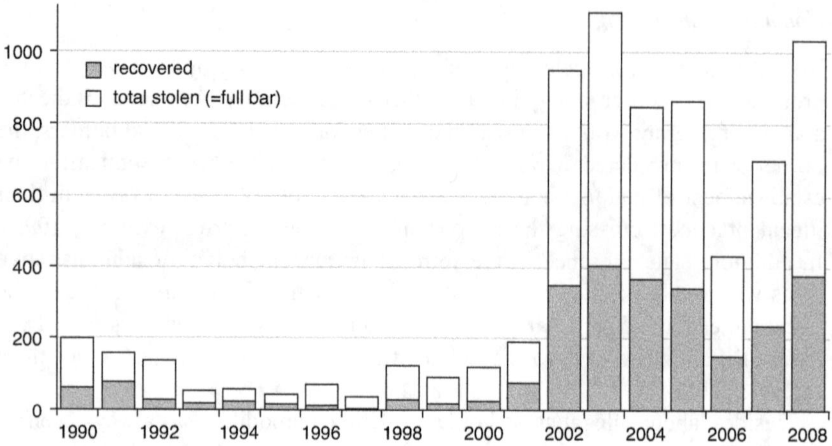

Figure 6.2 Number of cattle stolen and recovered: Southeast Lowveld, Zimbabwe, 1990–2008.

Source: Chikombedzi Police Stock Theft Records.

The post-2000 period saw a rapid rise in cattle theft and cross-border movements of cattle in the Southeast Lowveld (Figure 6.2). It was encouraged by three factors. First, the formal cattle market collapsed, as a result of fast-track land reform. Even in the 1990s, smallholder cattle keepers still relied for their sales on the nearby large-scale commercial enterprises trading large volumes. As the latter category fell away as a result of the land reform programme, the formal marketing structure also collapsed.

Second, the violence and chaos surrounding the land reform programme resulted in an upsurge in cattle thefts. Both large-scale commercial enterprises and smallholders saw substantial numbers being 'confiscated' by land reform beneficiaries, and by cattle rustlers who sell cattle illegally across the border in Mozambique. Simultaneously, many cattle keepers faced with imminent eviction and/or violence also tried to move their cattle illegally to Mozambique (Mavedzenge *et al.*, 2008, p. 629).

Third, rapid economic decline accompanied by commodity shortages and hyperinflation undermined the sale of cattle within Zimbabwe in favour of selling on the Mozambican side of the border. During the years 2002 to 2008, cattle were bought at prices between (US$500 and US$1000) in Mozambique as compared with an equivalent of US$100 in Zimbabwe. Cattle rustling thus became more lucrative, but also cattle owners preferred to sell across the border. Reports of theft thus started to include illegal trade by cattle owners themselves, as they provided reporters with permits to cross the border to recover their 'stolen' livestock and buy other necessities no longer available or affordable in crisis-ridden Zimbabwe.

Although foreigners, and Mozambicans in particular, are often blamed for the increase in cattle theft and rustling, the post-2000 political and economic situation provided clear incentives for local people to sell illegally across the border.

Close historical and socio-cultural ties between the Shangani peoples on either side of the international border make it also difficult to establish who is foreign or local.

Conflicting valuations, contradictions and the creation of ambiguity

Because people and institutions understand the edge differently, it easily becomes a place for divergent programmes. In the cases below, we look at how ministries in the same government pushed different and conflicting initiatives arising from differing conceptions of the edge.

Eco-tourism vs. commercial agriculture: government ministries at loggerheads in Mozambique

The governing body for conservation areas in Mozambique lies within the Ministry of Tourism. The Limpopo National Park and the GLTFCA thus fall under the jurisdiction of this national ministry. African Wildlife Foundation (AWF), anticipating the influx of the tourists that the LNP was to bring to the area and regarding the place as fit for conservation, started plans in the late 1990s to set up community-based game reserves. The plans involved extended consultations with communities as required by the National Land Law and AWF took five years to get the necessary permission to begin fencing off the game reserve.

On the other hand, the Ministry of Agriculture developed its own vision of the area when the opportunity arose. It saw the area as fit for agricultural production when the international sugar-producing company, ProCana, wanted to invest in biofuel plantations in the area. During the five years that AWF and its partners were trying to set up the eco-tourism projects that would benefit local communities, the Ministry of Agriculture was busy extending a concession to ProCana. The concession, which would involve an investment of US$520 million, was issued by the provincial Department of Agriculture in less than a year. In issuing the concession, the department justified their actions by publicizing the economic benefits, such as employment, that the project promised and juxtaposing those with the 'scant' economic development that a conservation area could bring.

The concession area is situated within the GLTFCA, along the road to the gate of the new Limpopo National Park. It also overlapped with the land set aside for the resettlement of villages from the park area, principally the land for grazing. The two giant projects, the TFCA and the sugar company, and the ministries involved were set in conflict.

The Economic Council of the National Ministries was called to resolve the conflict between the Ministry of Agriculture and the Ministry of Tourism. In the end, it was decided that the grazing area for the resettled communities displaced by ProCana be compensated for by providing an alternative area of nearly 430 km^2 from the proposed TFCA. This left the idea of game reserves and tourist places unfeasible and the AWF pulled out. A year later, the ProCana project collapsed and the planned development never took place.

For agriculture or wildlife? CAMPFIRE and MZRDP in Zimbabwe

The Department of National Parks and Wild Life Management (DNPWLM) saw the Dande area as a rich wildlife area. The department therefore accorded Guruve district authority status to operate the CAMPFIRE programme. The programme aimed at developing the Dande area through conservation and sustainable exploitation of wildlife and other natural resources. Since the inception of the programme in 1989, various wards, or producer communities as they were envisaged in the founding document (Martin, 1986), have benefited from commercial exploitation of wildlife, earning revenue from wildlife through safari hunting. This includes Dande residents who had abundant wildlife especially in the northern part of Guruve. At the same time, the Ministry of Agriculture and Rural Development saw the Dande area as a resettlement and agricultural zone. In this connection, it sought the support from the Food and Agriculture Organization (FAO) to implement the MZRDP based on agricultural intensification and infrastructure development.

Thus a conflict emerged between the two big projects implemented by two different government agencies. The situation was made worse by subsequent developments in the area. Originally, MZRDP was supposed to be limited to southern Dande, leaving the northern part to wildlife and natural vegetation. Due to shortage of arable land in the southern part to resettle several thousands of households who had not been considered in the earlier plan, the MZRDP was then also implemented in the northern part, effectively conflicting with the CAMPFIRE programme (Derman, 1993). The DNPWLM and its local government partners such as the Guruve Rural District Council opposed the transformation of the wildlife-rich landscape that could benefit local people. For their part, FAO and the resettlement authorities reviled CAMPFIRE for interfering with more intensive utilization of the land for the benefit of households. Ideas about what the edge was for had given rise to conflict between projects and relevant government departments.

Another example of this conflict comes from Binga District where the DNPWLM and the Binga Rural District Council, the local authority, joined together to implement CAMPFIRE in the south-eastern part of the country, which they argued could not sustain agriculture at least in the long term (Dzingirai, 2003). This area was divided into two hunting concessions allocated to two different safari companies. Further down, the RDCs carved out a third concession for photographic safaris. The area was wild, and therefore appeared to fit these activities.

But, while the DNPWLM and the Binga Rural District Council were busy developing and implementing plans, the Ministry of Agriculture had its own plans organized around settlement of the district by ethnic groups. For the Ministry of Agriculture, Binga was meant for cotton farming and livestock production. To facilitate settlement and crop production, the ministry set up various collections depots for both cattle and agricultural commodities in southern Binga. Further, the ministry provided credit facilities to the cotton producers, which was supplemented by development of extension support to the many illegal settlers in Binga. Finally, the Department of Veterinary Services stepped up its efforts, developing infrastructures such as fences, dip tanks and cattle sale pens along paths and tracks that dissect the forests of southern Binga. Private cattle buyers were also introduced in the area, buying the cattle on behalf of the CSC, a state company whose operations fall under the Ministry of Agriculture. The latter had contractual obligation to supply beef to the EU, which was a source of foreign currency to the country. Even though beef production was important to the country, Binga Rural District Council, which was now benefiting from CAMPFIRE through commercial exploitation of wildlife, was opposed to it. From the 1990s towards the early 2000s, the local authority sought for ways to reverse or at least stop the illegal migration facilitated by government departments. Further details regarding this story, especially the various steps taken by Binga Rural District Council to reverse the occupation of wildlife land by migrants, are not important. The point is that various authorities had different ideas about the edge and the nature of development activities fitting it.

Conclusion

The purpose of this chapter was to analyse the manner in which the edge is valued, the basis of those valuations, and their implications. The key observations we make are that the edge is not only valued differently by those located or interested in it, but most importantly, these valuations are not necessarily linked to its proximity to conservation areas.

The multiple and contesting valuations of the edge mean a number of things. First, the edge gives rise to a mosaic of projects and practices – some legal, others not. Second, what the edge is and should be remains ambiguous, and thus it attracts more plans and projects from government or non-governmental organizations.

The suggestion from this chapter is that such projects will add to the complexity of land use practices on the edge. They will contribute to a landscape with no single direction of development. The condition of marginality persists, attracting even more competition on what the place should be and for whose benefit. This is, perhaps, the curse of the edge. How the edge can break away from this curse or dilemma and how it can become the focus of a shared vision across space, time and boundaries is a matter that warrants attention to theory.

Acknowledgements

The authors thank David Hughes, Marja Spierenburg and Harry Wels for comments on an earlier version of the chapter.

Note

1 The commonly used agro-ecological classification of Zimbabwe's land (Vincent and Thomas, 1960) is illuminating in this respect. Edge areas are generally located in Natural Regions IV and V that were considered unsuitable for crop farming.

References

Alexander, J., McGregor, J. and Ranger, T. (2000) *Violence and Memory: One Hundred Years in the Dark Forests of Matabeleland*, James Currey, London

Andersson, J.A. (2002) 'Administrators' knowledge and state control in colonial Zimbabwe: the invention of the rural-urban divide in Buhera district', 1912–80', *Journal of African History*, vol. 43, no. 1, pp. 119–143

Andersson, J.A. (2006) 'Informal moves, informal markets: international migrants and traders from Mzimba District, Malawi', *African Affairs*, vol. 105, no. 420, pp. 375–397

Aubin, S. (ed.) (1997) *De la Terre, des Animaux et des Hommes, Diagnostic d'un Système Agraire dans la Moyenne Vallée du Zambèze*, DAT thesis, CNEARC, DAA Ensam, Montpellier

Bannerman, J. (1978) 'Towards a history of the Hlengwe people of the south of Rhodesia', *NADA*, vol. 11, no. 5, pp. 483–496

Baudron, F., Corbeels, M., Andersson, J.A., Sibanda, M. and Giller, K.E. (2011) 'Delineating the drivers of waning wildlife habitat: the predominance of cotton farming on the fringe of protected areas in the Mid-Zambezi Valley, Zimbabwe', *Biological Conservation*, no. 144, pp. 1481–1493

Biodiversity Project (2002) *Mankind and the Animal in the Mid Zambezi Valley, Zimbabwe*, CIRAD, Montpellier

Carruthers, J. (2009) 'National parks in South Africa', in H. Suich and B. Child (eds), *Evolution and Innovation in Wildlife Conservation: Parks and Games Ranches to Transfrontier Conservation Areas*, Earthscan, London, pp. 35–49

CCJP (1997) 'Gukurahundi: report on the Disturbances in Matabeleland and Midlands 1980–1988,' Columbia University Press, New York

Child, G.F.T. and Riney, T. (1987) 'Tsetse control hunting in Zimbabwe, 1919–1958', *Zambezia*, vol. 14, no. 1, pp. 11–71

Cumming, D.H.M. and Lynam, T. (1997) 'Land use changes, wildlife conservation and utilization and the sustainability of agro-ecosystems in the Zambezi Valley', WWF Multispecies Project, Harare

Derman, B. (1993) 'Recreating common property management: government projects and land use policy in the Mid Zambezi Valley, Zimbabwe', Occasional Paper, Centre for Applied Social Sciences, University of Zimbabwe, Harare

Duffy, K. and Slotow, R. (2006) 'African elephant home range and habitat selection in Pongola Game Reserve, South Africa', *African Zoology*, vol. 41, no. 1, pp. 147–149

Dzingirai, V. (1998) 'Human migration and natural resources in Zimbabwe's communal lands: the case of Binga, Zambezi Valley', Ph.D. thesis, Centre for Applied Social Sciences, University of Zimbabwe, Harare

Dzingirai, V. (2003) 'The new scramble for the African countryside', *Development and Change*, vol. 34, no. 2, pp. 243–264

Fabricius, C. and Wet, C. de (2002) 'The influence of forced removals and land restitution on conservation in South Africa', in D. Chatty and M. Colchester (eds), *Conservation and Mobile Indigenous Peoples: Displacement, Forced Resettlement and Sustainable Development*, Berghahn Books, Oxford

Floyd, B.N. (1962) 'Land apportionment in Southern Rhodesia,' *Geographical Review*, vol. 52, no. 4, pp. 566–582

Harries, P. (1994) *Work, Culture and Identity: Migrant Labourers in Mozambique and South Africa c. 1860–1910*, James Currey, London

Ivy, P. (1998) 'Land use in Ward 10 and evaluation of carrying capacity in Ward 2, 3, 4, 5, 9 and 10, Dande Communal Land, Guruve District, Zimbabwe', Regional Tsetse and Trypanosomiasis Control Programme, Harare

Jager, P. de (1988) 'Environmental degradation in communal land', MSc thesis, Department of Biological Sciences, University of Zimbabwe, Harare

Jeeves, A. (1985) *Migrant Labour in South Africa's Mining Economy: The Struggle for the Gold*, McGill-Queen's University Press, Montreal

Lan, D. (1985) *Guns and Rain: Guerrillas and Spirit Mediums in Zimbabwe*, Zimbabwe Publishing House, Harare

Lundstrum, E.M. (2007) 'The making and unmaking of sovereign territory: from colonial extraction to postcolonial conservation in Mozambique's Massingir region', Ph.D. thesis, University of Minnesota

Marongwe, N. (2007) 'Redistributive land reform and poverty reduction in Zimbabwe', Working paper. Available online: www.lalr.org.za/zimbabwe/redistributive-land-reform-and-poverty-reduction-in-zimbabwe/view (accessed 11 November 2011)

Martin, J. and Johnson, P. (1981) *The Struggle for Zimbabwe*, Faber and Faber, London

Martin, R.D. (1986) 'The Communal Areas Management Programme For Indigenous Resources', Department of National Parks and Wild Life Management, Harare

Mavedzenge, B.Z., Mahenehene, J., Murimbarimba, F., Scoones, I. and Wolmer, W. (2008) 'The dynamics of real markets: cattle in southern Zimbabwe following land reform', *Development and Change*, vol. 39, no. 4, pp. 613–639

Mtisi, J., Nyakudya, M. and Barnes, T. (2009) 'War in Rhodesia', in B. Raftopoulos and A. Mlambo (eds), *Becoming Zimbabwe: A History from the Pre-Colonial Period to 2008*, Weaver Press, Harare, pp. 141–166

Murray, M. (1995) 'Blackbirding at Crooks-Corner: illicit labour recruiting in the northeastern Transvaal, 1910–1940', *Journal of Southern African Studies*, vol. 21, no. 3, pp. 373–397

Muzondidya, J. (2009) 'From buoyancy to crisis, 1980–1997', in B. Raftopoulos and A. Mlambo (eds), *Becoming Zimbabwe: A History from the Pre-Colonial Period to 2008*, Weaver Press, Harare, pp. 167–200

Nemarundwe, N. (2005) 'The performance of CBNRM in the face of socio-political dynamism: a case study of CAMPFIRE in Masoka, Zimbabwe', in V. Dzingirai and C. Breen (eds), *Confronting the Crisis in Community Conservation: Case Studies From Southern Africa*, CEAD, University of KwaZulu-Natal, South Africa

Newitt, M.D.D. (1995) *A History of Mozambique*, Indiana University Press, Bloomington

Niehaus, I. (2002) 'Ethnicity and the boundaries of belonging: reconfiguring Shangaan identity in the South African lowveld', *African Affairs*, vol. 101, no. 405, pp. 557–583

Nkomo, J.N. (1984) *The Story of My Life*, Methuen, London.

Nyambara, P.S. (2001) 'Immigrants, "traditional" leaders and the Rhodesian state: the power of "Communal" land tenure and the politics of land acquisition in Gokwe, Zimbabwe, 1963–1979', *Journal of Southern African Studies*, vol. 27, no. 4, pp. 771–791

Palmer, R. (1977) *Land and Racial Domination in Rhodesia*, Heinemann, London

Paton, B. (1995) *Labour Export Policy in the Development of Southern Africa*, University of Zimbabwe Publishers, Harare

Phimister, I.R. (1978) 'Meat and monopolies: beef cattle in Southern Rhodesia, 1890–1938', *Journal of African History*, vol. 19, no. 3, pp. 391–414

Phimister, I. (2008) 'The making and meanings of the massacres in Matabeleland', *Development Dialogue*, no. 50, pp. 197–218

Pollock, J.N. (1991) *The Angwa-Manyame Large-scale Target Trial Project*, Vakakis International S.A., Athens

Reid, M.G. (1976) 'Progress in cotton production in the Tribal Trust Lands', *Rhodesia Science News*, vol. 10, no. 10, pp. 255–256

Roder, W. (1965) 'The Sabi Valley irrigation projects', Department of Geography Research Paper No. 99, University of Chicago, Chicago

RTTCP (1995) *Mid-term Evaluation of Phase 2 of the Regional Tsetse and Trypanosomiasis Control Programme of Malawi, Mozambique, Zambia and Zimbabwe, including Namibia (Final report Annexe E: Zimbabwe)*, Regional Tsetse and Trypanosomiasis Control Programme, Harare

Saunders, C. (1989) *Murray MacDougall and the Story of Triangle*, Triangle Limited, Zimbabwe

Spierenburg, M. (1995) 'The role of the Mhondoro cult in the struggle for control over land in Dande (northern Zimbabwe): social commentaries and the influence of adherents', Centre for Applied Social Sciences, University of Zimbabwe, Harare

Vincent, V. and Thomas, R.G. (1960) *An Agro-ecological Survey of Southern Rhodesia: Part I Agro-ecological Survey*, Government Printers, Salisbury, Rhodesia

Weinrich, A.K.H. (1977) 'Strategic resettlement in Rhodesia', *Journal of Southern African Studies*, vol. 3, no. 2, pp. 207–229

Wolmer, W. (2003) 'Transboundary conservation: the politics of ecological integrity in the Great Limpopo Transfrontier Park', *Journal of Southern African Studies*, vol. 29, no. 3, pp. 261–278

Wolpe, H. (1972) 'Capitalism and cheap labour-power in South Africa: from segregation to apartheid', *Economy and Society*, vol. 1, no. 4, pp. 425

Worby, E. (2000) 'Discipline without oppression: sequence, timing and marginality in Southern Rhodesia's post-war development regime', *Journal of African History*, vol. 41, no. 1, pp. 101–125

Wright, A. (1972) *Valley of the Ironwoods: A Personal Record of Ten Years Served as District Commissioner in Rhodesia's Largest Administrative Area, Nuanetsi, in the South-Eastern Lowveld*, T.V. Bulpin, Cape Town

7 Resource gradients and movements across the edge of transfrontier parks

Amon Murwira, Michel de Garine-Wichatitsky, Fadzai Zengeya, Xavier Poshiwa, Steven Matema, Alexandre Caron, Chloé Guerbois, Eléonore Hellard and Hervé Fritz

Introduction

Boundaries of protected areas in Southern Africa have often been drawn arbitrarily, following geographic features such as rivers. They were also drawn on borders between states, on demarcations between public/private properties or sanitary cordons between disease-infested/disease-free areas (see Chapter 3). The designation of protected area boundaries rarely took into consideration the social and political context of the local communities, which later resulted in customary claims on land designated as wildlife sanctuaries. The Makuleke clan is an example of a community whose social, economic and political structures were strongly affected as a result of being removed from the Pafuri Triangle in 1969 (Friedman, 2005).[1]

Although boundaries of protected areas were not designed according to ecological attributes, the areas on either side of the boundary are now often characterized by marked discrepancies in the abundance of (natural) resources. Thus, the boundaries of protected areas are important structural and functional components of landscape mosaics, separating contrasting elements and modulating the connecting flows (Cadenasso *et al.*, 2003). For example, the establishment of the boundary of Gonarezhou National Park in 1975 was a measure to control wildlife movements, in particular to prevent contacts between wild herbivores and livestock. As a result, the park boundary encompassed some of the grazing areas used by the local people, especially in the southern region of the park. The establishment of the park resulted in a steep resource gradient with grazing resources for livestock being more abundant in the conservation area compared with the surrounding communal land, especially during the end of the dry season. The existence of steep resource gradients has often prompted the movements of livestock and wildlife, as well as people, across boundaries that separate farming and wildlife activities, resulting in human–wildlife conflicts (see Chapter 8).

In this chapter, we explore the hypothesis that gradients of natural resources are a major determinant of movements of wildlife, livestock and people, across the edge of protected areas located inside transfrontier conservation areas (TFCAs) in Southern Africa. Understanding the mechanisms that prompt crossing of the edge provides insight into possible interventions that will promote sustainable utilization of resources at the edge. We start by discussing concepts of ecological 'trade-offs' that are helpful in explaining how access to food resources may lead to movements of wild animals across the boundaries of protected areas. We discuss to what extent such mechanisms are relevant to explain movements of domestic animals and people living at the periphery of protected areas. Using case studies from Gonarezhou National Park (GNP) and Hwange National Park (HNP) in Zimbabwe, and to a lesser extent Kruger National Park (KNP) in South Africa, we first describe the resource gradients between protected areas and neighbouring communal lands. We then demonstrate how the spatial and temporal variability in the availability of food resources (grazing or prey) can lead to movements of wild herbivores and wild carnivores outside protected areas, resulting in human–wildlife conflicts, which are discussed in Chapter 8. Similarly, we analyse the movements of cattle inside national parks in an attempt to evaluate the relative importance of human factors (owners' and herders' decisions), as well as ecological gradients in influencing these movements. Finally, using a case study on bushmeat poaching in Zimbabwe, we illustrate the complexity of people's decisions to cross the edge, suggesting that the proximate factor may be a response to resource gradients (availability of game animals) although the ultimate reasons are many, including a complex interplay of cultural, political as well as symbolic reasons.

Factors driving animal movements

Movements by unattended animals, both wild and domestic, rely on ecological trade-offs along a gradient of resources. However, some of the costs associated with crossing the edge of protected areas may either be direct or indirect, and may have important management implications (see Chapter 8). Here we present ecological concepts that help to explain movements of wild and domestic species, even though the latter are also influenced by husbandry practices.

Several factors can lead animals to leave the relative safety of their core home range area. These can broadly be grouped into three categories: (1) the ecological conditions are more attractive outside (food, water, less competition or disturbance, better cover, etc.) their core area; (2) the ecological conditions are no longer viable (e.g. lack of food resources or unbearable disturbances), so that animals are pushed out of their core area; and (3) the animals follow short-term variation in resource abundance (e.g. cycling grazing or water commuting behaviour) or long-term variation in resource abundance (e.g. seasonal migration) movement routes. The outside of the core area is part of the home range; it has complementary resources (in space and time) that are used in addition to those found in the core area. None of these factors is exclusive of another, but reasons (1) and (3) may result in an 'ecological trap', i.e. organisms become 'trapped' by their innate behavioural responses to the cues across the boundary and experience reduced survival or reproduction success by moving (e.g. Kokko and Sutherland, 2001; Schlaepfer *et al.*, 2002). For instance, the provision of water outside the protected area could create a more attractive habitat in an arid landscape, but the risk associated (being shot, or snared) may be greater than the benefit of accessible water. At the population level, the movements originating from (1) and (3) will lead to the establishment of 'source–sink' systems, among which those resulting from ecological traps have the greatest impact (Kristan, 2003; Battin, 2004). 'Ecological traps' are often induced by human modification of the environment and generally occur in a shorter time span than natural environmental changes. For that reason, domestic stock may respond similarly to environmental changes, however herding often serves to cancel the 'source–sink' effect. For wild species the 'source–sink' phenomenon may have multiple effects. Indeed, if the magnitude of the change exceeds the range of values normally encountered, some populations may be unable to survive the new circumstances. Nevertheless, if properly managed, these 'source–sink' systems may be a sustainable way to exploit natural resources (Novaro *et al.*, 2000) and hence favour coexistence between protected areas and their neighbouring communities. Transboundary movements are often associated with (3), and are often seasonal and long-distance movements.

In the case of reason (2), disturbances pushing animals out of their territory may be induced by humans (e.g. poaching or grazing management), by animals (e.g. interference competition among carnivores or crop protection) or by climate (e.g. drought). Because they affect the viability of the core area, these cannot be turned into viable source–sink systems without major intervention. Once again, these effects on domestic stock are often milder than for wild species.

Factors driving movement of people

Voluntary movements by people across the edge are triggered by various factors such as disruptive political, environmental, economic and social conditions (Datta, 2003). However, Datta (2003) further elaborates that at the individual level, factors such as degree of skills, family circumstances and personality also play a significant role in motivating an individual to move. Our understanding of the factors affecting movement from one place to another is based on the premise that an individual realizes benefits accrued by the move (Rabianski, 1971). For instance, environmental adversity such as drought results in food insecurity and is a significant driver of movement across the edge, leading to an increase in activities such as bushmeat hunting (Bulte and Rondeau, 2007). Voluntary movements across the edge thus become a coping strategy to offset economic hardships and food insecurity during difficult periods.

However, crossing the edge does not always result from rational decisions to optimize economic benefits. It may also be triggered by a need to visit communities on the other side or to access ancestral burial and sacred sites in the protected areas. The head of the Nemadire clan, which was driven out in the early 1960s of the area that is now included in Save Valley Conservancy in Zimbabwe, argues that unless access to ancestral burial sites located inside the conservancy is granted, the aggrieved ancestral spirits will continue to possess the young men and lead them into the conservancy to hunt (Matema, 2010).

Resource gradients across the edge of protected areas

Spatial variation in ecological and disturbance factors within landscapes results in a heterogeneous distribution of natural resources (de Knegt *et al.*, 2007). Protected areas often act as reserves of grazing, water and other natural products such as thatch grass, fuelwood and construction material for the people living at the periphery of protected areas. Thus a gradient of resources exists across the edge of protected areas, which varies in nature both spatially and temporally.

Grazing and browsing resources

Data from ungrazed exclosures and grazed control plots in communal lands, small-scale commercial land and the GNP show significant differences in standing grass biomass between the three areas during the rainy season for the grazed control plots (Figure 7.1), the least standing biomass being found in the communal land. No significant differences were recorded in the ungrazed exclosures between the three areas, indicating that the grass productivity during the growing season was similar. As the rainy season progresses, the decrease in the available standing grass biomass in the communal land creates a steep gradient between the communal land, the GNP and the small-scale commercial farming land. As grazing resources become relatively more abundant in small-scale commercial farming areas and the GNP compared with the communal lands during the beginning of the dry season, this attracts cattle to move into the areas where grazing resources are relatively more abundant.

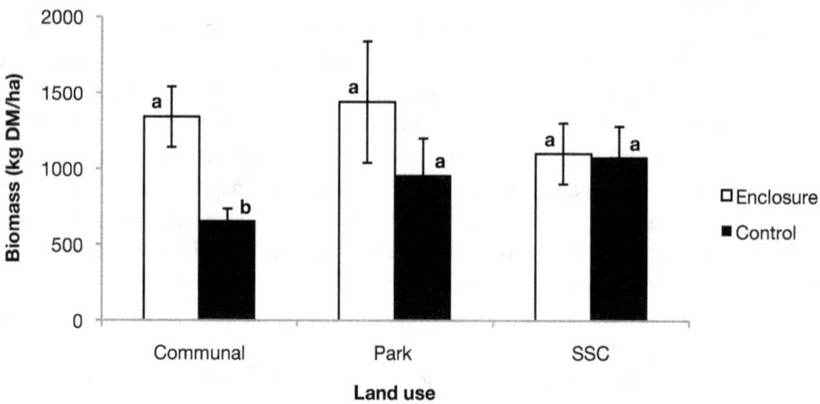

Figure 7.1 Gradient of grazing resources according to land use in the GLTFCA. Standing grass biomass (kg DM/ha) in communal, park and small-scale commercial areas during 2008–2009 rainy season in exclosures (open bars) and control plots (black bars). Error bars represent 95% confidence intervals and bars with the same letter are not significantly different (LSD test, $P > 0.05$).

The steep gradient of grazing resources observed between GNP and the neighbouring communal land results from different management in these areas. Within conservation areas predation, illegal hunting and snaring and occasional authorized hunts regulate the wild herbivore populations, in addition to the effects of availability of feed. In communal lands, a farmer can own as many animals as he wants, and can increase his herd size without considering the availability of grazing resources. The human population, and the number of cattle, has increased drastically over the past century. In Matibi No. 2 communal land, Cumming (2005) estimated that the human population has multiplied 25-fold in the past 80 years from 2,000 to 50,000, with a resulting decrease of land availability from 650 ha per household in 1920 to 35 ha per household in 2000. This increase in demographic pressure, which has been accompanied by an increase in cattle numbers explains the current grazing pressure within the communal lands compared with the GNP and the small-scale commercial areas.

Abundance of wild and domestic prey

The abundance of prey is arguably one of the main factors inducing the movements of wild predators. The abundance of prey differs significantly across the edges of protected areas, in terms of species composition and relative densities. Densities of both wild and domestic herbivores were estimated using road counts inside GNP and in the neighbouring Malipati Communal Land in 2008 and 2009 (de Garine-Wichatitsky *et al.*, unpublished data). During the dry season (August 2008), local apparent densities in the vicinity of Malipati village adjacent to GNP were 30.9 cattle/km^2 [10.7–89.7] and 42.0 goats/km^2 [12.9–136.4], whereas densities of wild herbivores in Mabalauta area of GNP during the same period ranged from 2.2 kudu/km^2 [0.8–5.9] to 14.8 impala/km^2 [4.7–46.8]. Overall, the density of herbivores was higher in the communal land than across the boundary in the protected area.

However, the existence of significant differences in the densities of herbivores does not necessarily mean that the numbers of prey available to predators follow the same gradient. Indeed, there is little doubt that the capacity to escape predation is higher for wild ungulates than domestic species. But livestock benefit from protection by herders. In the Malipati area, cattle, goats and donkeys are usually kept at night in kraals located near the houses of their owners to protect them against theft and predation, and only under exceptional circumstances do livestock remain in the bush overnight. The efficiency of the protection provided against predators is highly dependent on the design of the kraal and herding practices during the day.

Other natural resources

Building materials, such as wood poles or thatch grass, and firewood are often more abundant inside protected areas compared with neighbouring communal lands. The biomass of wood in miombo woodlands is usually significantly greater inside protected areas compared with the neighbouring communal lands where it is harvested for firewood or charcoal (e.g. Luoga *et al.*, 2002). However, conservation policies may sometimes result in the opposite effect, such as when abundant elephant populations significantly reduce woody cover inside protected areas (e.g. Mapaure and Campbell, 2002).

In the Southeast Lowveld of Zimbabwe, for instance, '*Musimbiti*' wood poles (*Spirostachys africana*) are highly valued for construction as the hard wood is resistant to termite attack. Local communities in Malipati and Chikombedzi communal lands acknowledge that this tree has been overharvested and that large diameter poles are mostly found nowadays inside GNP and Malipati Safari Area. In 2009, a formal permission was granted for collection of '*Musimbiti*' poles in the GNP to rehabilitate the infrastructures of dip tanks.

Wildlife movements at the edges of national parks

Despite having contributed greatly to biodiversity conservation, the creation of protected areas has also led to habitat compression by limiting home ranges and migration routes of wildlife. At the same time, the expansion of agricultural lands and increasing human population density closer to protected areas have also contributed to habitat compression. Increasing wildlife populations has led to competition for resources within the limited space of protected areas resulting in wildlife moving out of the protected areas in search of prey or grazing and browsing resources.

In areas around GNP, predation data show that a large proportion of households who lose livestock to lion live close to the park, especially where there is dense woody cover as represented by Region A in Figure 7.2. In Figure 7.2 Region B fewer livestock kills are experienced as a result of low woody cover and increased distance from the park. Although the reasons for lion movement out of the GNP to prey on livestock are unclear, it is likely that they do so because domestic stock are easy prey and livestock densities are relatively higher in the communal land. Studies in Asia revealed a similar trend, where areas closer to the conservation areas tend to be more associated with human–wildlife conflict than those further away (Watoa *et al.*, 2006; Thapa, 2010).

Further insights from HNP suggest that the increase in predator density within the park and increasing human and livestock populations on the border tend to attract predators into the surrounding communal lands. Here, a marked increase in lion and hyena populations, coupled with a decline of herbivores inside the park, seems to have pushed the carnivores out of the park in search of prey. Concurrently, the human population (almost doubled since 2000) as well as the livestock populations (more than doubled) has increased, thereby amplifying the chances of human–wildlife conflicts. Increased predator density and increased human population density have been cited as major drivers of human–predator conflict in many areas in Asia and Africa (Wang and Macdonald, 2006).

Between October 2008 and May 2009, interviews were carried out in four villages in the northern part of HNP to record hyena attacks on livestock and their

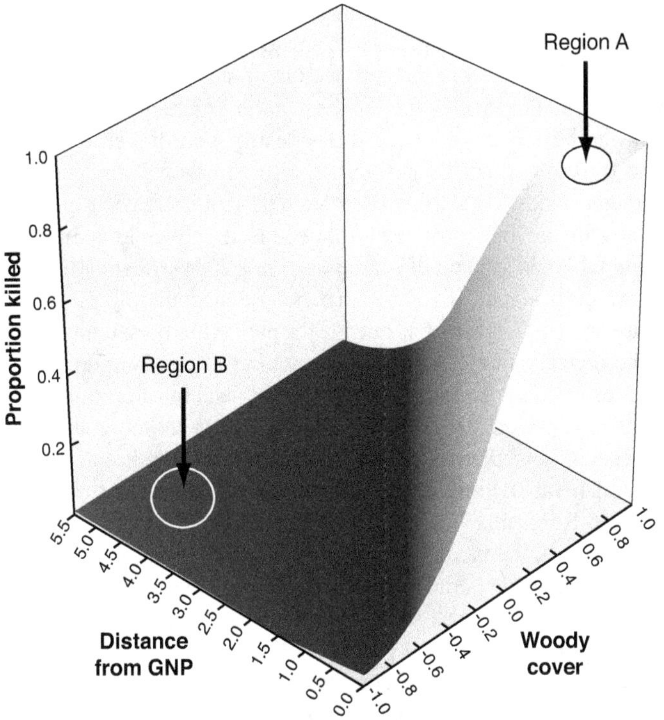

Figure 7.2 Lion predation on livestock in areas adjacent to Gonarezhou National Park, Zimbabwe. The graph shows the proportion of livestock (cattle, goats and donkeys) killed by lion as a function of distance (km) from GNP and amount of woody cover estimated using a Normalized Differene Vegetation Index (NDVI) (based on numbers recorded by Kuvaoga in 2008).

circumstances since 2007 and to investigate livestock management and agricultural practices. Among the 31 households visited, 20 households had livestock killed by spotted hyenas since 2007. In total 100 animals had been killed in 28 attacks, mostly goats (66.7 per cent), followed by cattle (23.3 per cent) and sheep (10 per cent). Most of the attacks occurred during the end of the dry season (September–October, 29.6 per cent) and the rainy season (November–May, 59.3 per cent), when people are working in the fields. At the end of the dry season grazing is sparse, leading the unherded livestock to move closer to the park inside the Sikumi Forest Area (adjacent to the park and the communal land). During the crop harvesting period, farmers push livestock away from their fields, resulting in the animals entering the Sikumi Forest Area, which increases the probability of attack by hyenas.

The consequences of predation on livestock by wild carnivores are significant (see Chapter 8). On the other hand, the consequences of predators crossing the edges of TFPs are also significant for biodiversity conservation. The endangered wild dogs (*Lycaon pictus*) population in Hwange illustrates the consequences of the 'ecological trap' described earlier in this chapter. Van der Meer (2011) documented the territorial drift of wild dog packs into an attractive sink located outside the park boundary, which resulted in a sharp decline of survival. Between 1989 and 2009, 327 African wild dogs were reported dead; almost three-quarters of these deaths occurred in the buffer zone outside the park and were linked to human activities (i.e. road traffic, snares). In addition to carnivore movements into the surrounding communal lands, data from HNP suggest that increasing elephant population densities inside the park lead to increased movements into the surrounding communal lands (Chamaillé-Jammes *et al.*, 2008). Results from 66 interviews in the Magoli-Jwape area show that people surrounding HNP also perceive this increase in elephant densities outside the park (Guerbois, unpublished data). However, the increase in elephant movements into the surrounding communal lands coincides with the period of crop harvesting and most crop raids are recorded during this time period. Close to harvest time, crops are more nutritious than any natural vegetation (Sukumar, 1990; Chiyo *et al.*, 2005). Elephants can destroy large tracts of fields overnight, representing up to 70 per cent of family production (Bulte and Rondeau, 2007). In related studies, differences in food quality between the protected area and the surrounding communal land are also known to drive elephants outside of the KNP. There is a growing body of evidence that elephant crossings are a response to the availability of marula (*Sclerocarya birrea*) fruits outside the KNP in summer (Gadd, 2002; Jori *et al.*, 2009). In some instances, local farmers have had to abandon their irrigation plots as is the case of Gumera and Chisavanye in Chipinge District, Zimbabwe, due to crop raiding by elephants and destruction of irrigation infrastructure (Interview, Ward 22 Councillor, Chipinge, August 2006).

Cattle movements at the edge

The grazing and browsing resource gradient that exists between communal lands and the protected areas of Southern Africa plays a role in influencing cattle farmers' grazing strategies, particularly during the dry season when grazing is scarce.

Cattle observations by game rangers in GNP, Zimbabwe show that people graze their cattle in the park towards the end of the dry season, between September and November (Figure 7.3). An average of 167 cattle was seen in the park at the end of the dry season in November compared with only 10 during the wet season in January. Interestingly, this is in contrast with the movement pattern described above for the villages located in the northern part of HNP, as their cattle were apparently entering the protected area more often during the early rainy season when they were pushed away from the field during the cropping season. The incidence of fines to cattle owners whose cattle were found in the GNP follows the same trend as cattle numbers found in the park (Figure 7.3).

Data from global positioning system (GPS)-collared cattle herds for September 2008 also show clearly that cattle graze in the GNP during the dry season (Figure 7.4). Allowing cattle to graze in the protected area can be understood as a deliberate response to a resource gradient. Such herding behaviour is not the

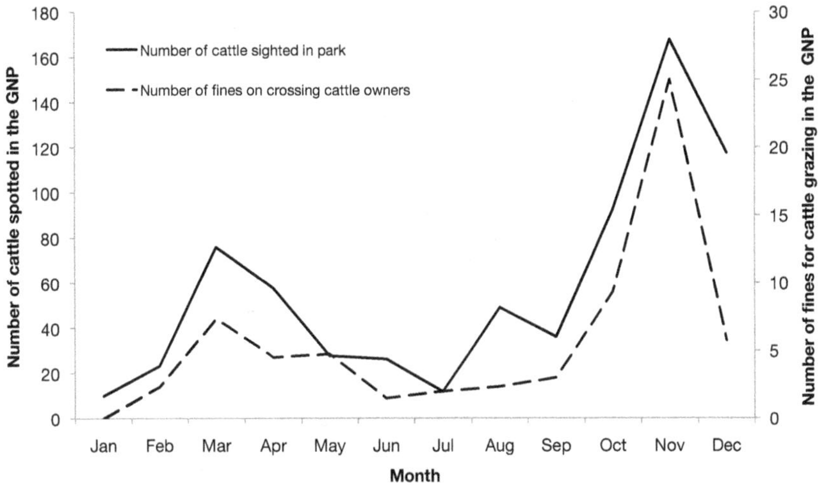

Figure 7.3 Average number of cattle reported inside Gonarezhou National Park, and number of fines issued. Data collated from patrol reports and monthly records by NPWMA for the period 2002–2009.

only solution to grazing scarcity in the communal lands. To the contrary, on the edge of GNP, communal land cattle farmers make arrangements with small-scale commercial farmers in the Gonakudzingwa area, who generally have substantial grazing resources throughout the year.

Furthermore, when asked about grazing issues and coping strategies, cattle keepers at the Malipati Animal Health and Management Centre in the Southeast Lowveld of Zimbabwe (Caron *et al.*, unpublished data 2007–2008), 30.6 per cent of the farmers living close to the park acknowledged sending their cattle in the GNP during the dry season. In the HNP system, the use of grazing land and peripheral permanent waterholes within protected areas (mostly Forestry land) was negotiated by the communities during extreme dry spells of the early 1990s (up to 3 km inside the park). Livestock keepers continue to access these grazing areas and waterholes despite repeated warnings from the Forestry Commission, and have increased their use in recent years (based on evidence from monthly stakeholders' meetings). The population of livestock has more than doubled in the last ten years in the Dete-Mabale area (Department of Veterinary Services, unpublished records). Buying cattle has long been a main way of investing capital for people living in communal lands of Zimbabwe, but it is possible that this has increased in response to the economic crisis which affected the economy of the country during the last decade. Limited cattle trade possibilities in most areas bordering conservation areas classified as Foot and Mouth Disease infected areas compound the problem of increasing cattle populations. Overall, these observations suggest that cattle movements into the conservation areas are, among other reasons, a response by cattle farmers to cope with relative scarcity in grazing and browsing resources on their land, especially during dry periods.

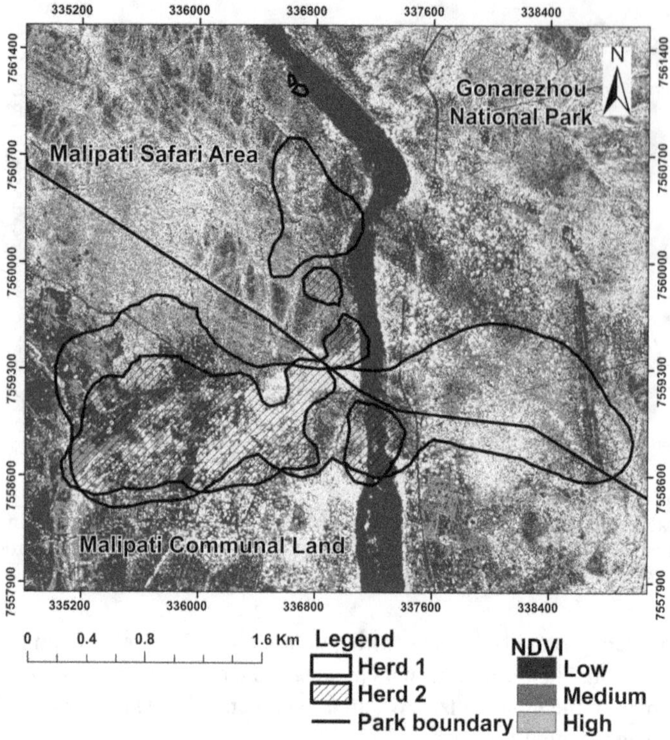

Figure 7.4 Home ranges of two cattle herds[2] and distribution of grazing and browsing resources at the periphery of Gonarezhou National Park in September 2008. The background image shows the spatial distribution of the normalized difference vegetation index (NDVI). High NDVI values indicate high vegetation cover (high grazing and browsing resource availability) while low NDVI values indicate low vegetation cover (less grazing and browsing resource availability).

People movements at the periphery of protected areas

Movement of people at the edges of protected areas has also been linked to access of non-timber forest products, which contribute substantially to local community incomes (Ninan, 2006). The Save Valley Conservancy (SVC) in the Southeast Lowveld of Zimbabwe provides a case study illustrating people movements across protected areas in relation to bushmeat hunting. The SVC made a transition from commercial ranching to wildlife ranching in 1992 after the former was no longer viable (Pole, 2006; Lindsey *et al.*, 2009; Matema, 2010).

Results of a survey carried out in 2006[3] provide an understanding of the importance and the reasons behind illegal wildlife hunting from the SVC by individuals from neighbouring rural communities. In a survey of illegal hunters ($n = 150$ hunters), the most frequently mentioned reasons for hunting in the protected areas were related to lack of income (52 per cent) or providing sufficient food (48 per cent) for family upkeep, although a significant number (11 per cent) also indicated hunting was a protest against the establishment of the conservancy.

Bushmeat hunting thus represents an important activity for some poor members of rural communities as a way to cope with economic hardship and food insecurity during difficult periods. In a separate but related study in communities around HNP, i.e. in the Dete area, 37 per cent of the 66 interviewed community members also stated that employment was the best way to reduce poaching. These findings point towards small-scale bushmeat hunting in the protected areas of TFCAs as a coping strategy against economic hardship, as opposed to large-scale poaching of highly-valued species such as rhinos and elephants, which are operations controlled by sophisticated criminal networks (TRAFFIC, 2011).

Discussion and conclusions

Our analysis indicates that the edge is a complex place where the greater availability of grazing and woodland products attracts local people, particularly during dry seasons. In addition, local people become involved in illegal hunting activities as a way of supplementing their income and enhancing their food security especially during lean times. Thus, crossing into the park plays a key role in the livelihoods of people living at the edge.

It is also clear that variation in space and time of the quality of food resources may drive wild animals outside of the protected areas. In the case of herbivores, such as elephants, this movement is driven by availability of crops in settled areas close to the park. In areas such as KNP, the availability of other resources of high quality such as the marula fruits, also leads elephants to move outside the park boundary. For predators, the high carnivore densities in the protected areas, coupled with the availability of easy prey such as cattle, donkeys and goats in the settled areas outside the park attracts hyena and lion.

The case of the SVC shows that the establishment of commercial farms during colonial days as well as the establishment of protected area boundaries took little consideration of the livelihood and cultural needs of neighbouring communities. Consequently conflicts related to the crossing of the SVC boundary are not solely driven by resource needs, but also by their historical use by local communities. When addressing resource use conflicts at the edge all of these ecological, historical and cultural factors must be considered to ensure sustainable wildlife conservation. Twinning development assistance is one option that has been deployed by global players to curtail bushmeat hunting but often this has failed partly due to poor targeting (Robinson and Bennett, 2002). However assessing employment needs, i.e. level of skills within communities, may help in designing employment opportunities for a majority of community members and reduce bushmeat hunting.

The consequences of movements across the edge are addressed in Chapter 8, but the existence of strong responses to resource gradients provides insights and tools for managing the edges. In other words, could managers manipulate resource gradients to control or regulate the crossing of the edge through the different management regimes that exist across the edge? Few attempts have been made to actively rely on the ecological and behavioural responses at the edge to generate management initiatives. As it is clear that the HNP and its periphery is greatly influenced by the availability of surface water (Chamaillé-Jammes *et al.*, 2007a, 2007b; Martin *et al.*, 2010), the management of water distribution may clearly help mitigate the movements of animals across the edge. Grazing and fire management of the buffer area (improving grass quality and regrowth) may allow for a better manipulation of the gradient, offering resource-rich habitat at the edge to both wild and domestic herbivores. This could reduce the frequency of both livestock and wildlife crossing of the edge, but may generate more contact between wild and domestic herbivores, which may not be desirable due to disease transmission. Disturbance shooting and hunting are well known to influence wildlife distribution, and in particular wildfowl game species (Madsen and Fox, 1995) and predation risk in general will shape habitat selection (Lima, 1998). These practices could offer sustainable management options (Treves, 2009), and enhance community support, especially if combined with the development of game meat production systems at the periphery of wildlife-rich protected areas.

Acknowledgements

The authors are grateful to Onisimo Mutanga and Harry van der Linde for reviewing this chapter. Original material and personal communications were kindly provided by Patience Zisadza, Ferran Jori, and Philip Kuvaoga. We are grateful to the National Parks and Wildlife Management Authority of Zimbabwe, whose staff have actively participated in the field surveys and provided useful information. We would like to express our sincere gratitude to the local communities and traditional leaders of the areas where the various surveys were conducted, with a very special thanks to the communities in Chiredzi District and Hwange District in Zimbabwe.

Notes

1 The Pafuri Triangle is found in the northern-most section of the KNP spanning the borders of northern South Africa, Zimbabwe and western Mozambique. The area is found between the confluence of Limpopo and Luvuvhu rivers and forms a distinct ecological region which supports a high level of biodiversity.

2 Twelve cattle herds were fitted with GPS collars, recording the GPS position of the animal every hour, during the period August–September 2008 in Malipati Communal Land in order to monitor the movement of cattle at the edge of the protected areas (Malipati Safari Area and the GNP). Distribution maps using data for two of the cattle herds illustrate the seasonal use of grazing and water resources inside the protected areas.

3 The survey was part of a study on land use, illegal bushmeat hunting and the future of wildlife in Zimbabwe from April 2006 to September 2006. Hunters, bushmeat buyers and SVC game scouts were interviewed to determine the dynamics and underlying causes of illegal hunting in and around the SVC. A total of 150 illegal hunters were interviewed in five sites in Buhera, Chiredzi, Bikita and Chipinge districts surrounding the SVC.

References

Battin, J. (2004) 'When good animals love bad habitats: ecological traps and the conservation of animal populations', *Conservation Biology*, vol. 18, pp. 1482–1491

Bulte, E. and Rondeau, D. (2007) 'Compensation for wildlife damages: habitat conversion, species preservation and local welfare', *Journal of Environmental Economics and Management*, vol. 54, pp. 311–322

Cadenasso, M.L., Pickett, S.T.A., Weathers, K.C., Bell, S.S., Benning, T.L., Carreiro, M.M. and Dawson, T.E. (2003) 'An interdisciplinary and synthetic approach to ecological boundaries', *BioScience*, vol. 53, no. 8, pp. 717–722

Chamaillé-Jammes, S., Fritz, H. and Murindagomo, F. (2007a) 'Climate-driven fluctuations in surface-water availability and the buffering role of artificial pumping in an African savanna: implications for herbivore dynamics', *Austral Ecology*, vol. 32, pp. 740–748

Chamaillé-Jammes, S., Valeix, M. and Fritz, H. (2007b) 'Managing heterogeneity in elephant distribution: interactions between elephant population density and surface-water availability', *Journal of Applied Ecology*, vol. 44, no. 3, pp. 625–633

Chamaillé-Jammes, S., Fritz, H., Valeix, M., Murindagomo, F. and Clobert, J. (2008) 'Resource variability, aggregation and density dependence: the local regulation of an African elephant population', *Journal of Animal Ecology*, vol. 77, pp. 135–144

Chiyo, P.I., Cochrane, E.P., Naughton, L. and Basuta, G.I. (2005) 'Temporal patterns of crop raiding by elephants: a response to changes in forage quality or crop availability?', *African Journal of Ecology*, vol. 43, pp. 48–55

Cumming, D. (2005) 'Wildlife, livestock and food security in the South East Lowveld of Zimbabwe', in S.A. Osofsky, S. Cleaveland, W.B. Karesh, M.D. Kock, P.J. Nyhus, L. Starr and A. Yang (eds), *Conservation and Development Interventions at the Wildlife/Livestock Interface: Implications for Wildlife, Livestock and Human Health*, IUCN, Gland

Datta, A. (2003) *Human Migration: A Social Phenomenon*, Mittal Publications, New Delhi, p. 176

Friedman, J. (2005) 'Winning isn't everything: what the Makuleke lost in the process of land restitution', B.A. dissertation, University of Chicago, Chicago

Gadd, M.E. (2002) 'The impact of elephants on the marula tree *Sclerocarya birrea*', *African Journal of Ecology*, vol. 40, pp. 328–336

Jori, F., Vosloo, W., Du Plessis, B., Bengis, R., Brahmbhatt, D., Gummow, B. and Thomson, G.R. (2009) 'A qualitative risk assessment of factors contributing to foot and mouth disease outbreaks in cattle along the western boundary of the Kruger National Park', *Revue Scientifique et Technique-OIE*, vol. 28, no. 3, pp. 917–933

Knegt, H.J. de, Hengeveld, G.M, Van Langevelde, F., De Boer, W.F. and Kirkman, K.P. (2007) 'Patch density determines movement patterns and foraging efficiency of large herbivores', *Behavioral Ecology*, vol. 18, pp. 1065–1072

Kokko, H. and Sutherland, W.J. (2001) 'Ecological traps in changing environments: ecological and evolutionary consequences of a behaviourally mediated Allee effect', *Evolutionary Ecology Research*, vol. 3, pp. 537–551

Kristan, W.B.I. (2003) 'The role of habitat selection behavior in population dynamics: source–sink systems and ecological traps', *Oikos*, vol. 103, pp. 457–468

Kuvaoga, P. (2008) 'Spatial distribution of wildlife–livestock conflict around Gonarezhou National Park', MSc dissertation, University of Zimbabwe, Harare

Lima, S.L. (1998) 'Nonlethal effects in the ecology of predator-prey interactions', *BioScience*, vol. 48, pp. 25–34

Lindsey, P.A., Romañach, S.S., Matema, S., Matema, C., Mupamhadzi, I. and Muvengwi, J. (2009) 'Dynamics and underlying causes of illegal bushmeat trade in Zimbabwe', *Oryx*, vol. 45, no. 1, pp. 84–95

Luoga, E., Witkowski, E. and Balkwill, K. (2002) 'Harvested and standing wood stocks in protected and communal miombo woodlands of eastern Tanzania', *Forest Ecology and Management*, vol. 164, no. 1–3, pp. 15–30

Madsen, J. and Fox, A.D. (1995) 'Impacts of hunting disturbance on waterbirds: a review', *Wildlife Biology*, vol. 1, pp. 193–207

Mapaure, I. and Campbell, B. (2002) 'Changes in miombo woodland cover in and around Sengwa Wildlife Research Area, Zimbabwe, in relation to elephants and fire', *African Journal of Ecology*, vol. 40, pp. 212–219

Martin, J., Chamaillé-Jammes, S., Nichols, J.D., Fritz, H., Hines, J.E. and Fonnesbeck, C. (2010) 'Modelling simultaneously habitat suitability, occupancy and abundance: case study of African elephants in Zimbabwe', *Ecological Applications*, vol. 20, pp. 1173–1182

Matema, S. (2010) *Conservancy-Edge Communities, Wildlife and Rural Livelihoods: The Dynamics and Underlying Causes of Illegal Hunting in Save Valley Conservancy*, Lambert Academic Publishing, Harare

Meer, E. van der (2011) 'Is the grass greener on the other side? Testing the ecological trap hypothesis for African wild dogs (*Lycaon pictus*) in and around Hwange National Park, Zimbabwe', Ph.D. dissertation, University Claude Bernard – Lyon1, Lyon

Ninan, K.N. (2006) 'Non timber forest products and biodiversity conservation: a study of tribals in a protected area in India', Proceedings of the 26th International Conference of Agricultural Economists, Gold Coast, Queensland, Australia, 12–18 August

Novaro, A.J., Redford, K.H. and Bodmer, R.E. (2000) 'Effect of hunting in source–sink systems in the Neotropics', *Conservation Biology*, vol. 14, pp. 713–721

Pole, A. (2006) 'Management plan for Save Valley Conservancy The Lowveld Conservancies: new opportunities for productive and sustainable land use', Report to the Save Valley Conservancies, Harare

Rabianski, J. (1971) 'Real earnings and human migration', *The Journal of Human Resources*, vol. 6, no. 2, pp. 185–192

Robinson, J.G. and Bennett, E.L. (2002) 'Will alleviating poverty solve the bushmeat crisis?', *Oryx*, vol. 36, no. 4, pp. 322–332

Schlaepfer, M.A., Runge, M.C. and Sherman, P.W. (2002) 'Ecological and evolutionary traps', *Trends in Ecology and Evolution*, vol. 17, no. 10, pp. 474–480

Sukumar, R. (1990) 'Ecology of the Asian elephant in southern India. II. Feeding habits and crop raiding patterns', *Journal of Tropical Ecology*, vol. 6, pp. 33–53

Thapa, S. (2010) 'Effectiveness of crop protection methods against wildlife damage: a case study of two villages at Bardia National Park, Nepal', *Crop Protection*, vol. 29, no. 11, pp. 1297–1304

TRAFFIC (2011) 'Rhino poaching in South Africa reaches new high'. Available online: www.traffic.org/home/2011/1/12/rhino-poaching-in-south-africa-reaches-new-high.html (accessed 29 August 2011)

Treves, A. (2009) 'Hunting for large carnivore conservation', *Journal of Applied Ecology*, vol. 46, pp. 1350–1356

Wang, S.W. and Macdonald, D.W. (2006) 'Livestock predation by carnivores in Jigme Singye Wangchuck National Park, Bhutan', *Biological Conservation*, vol. 129, pp. 558–565

Watoa, Y.A., Wahungu, G.M. and Okello, M.M. (2006) 'Correlates of wildlife snaring patterns in Tsavo West National Park, Kenya', *Biological Conservation*, vol. 132, pp. 500–509

8 Consequences of animals crossing the edges of transfrontier parks

Michel de Garine-Wichatitsky, Hervé Fritz, Petronella Chaminuka, Alexandre Caron, Chloé Guerbois, Davies M. Pfukenyi, Collen Matema, Ferran Jori and Amon Murwira

Introduction

Poor rural households in developing countries often survive on a combination of on-farm and off-farm activities. Chapter 4 described such diversified livelihood pursuits adopted by people living at the periphery of protected areas in Southern Africa. Natural resources may act as a buffer during periods of socio-economic instability or absorb environmental shocks (Paumgarten, 2006). Bush food products, for instance, play a major role in societies that experience seasonal food shortages (de Garine, 1993, 2005). Accessing water and grazing resources during droughts is a matter of survival for livestock in the semi-arid ecosystems of Southern Africa. Competition with wildlife for these scarce resources can exacerbate such scarcities for people living on the edge of protected areas. At the same time, the protected area may provide an alternative source of grazing, and thus a 'buffer' during these difficult periods.

Chapter 7 described how natural resources are unevenly distributed across the boundaries of protected areas. These resource gradients induce movements of livestock and wildlife, and to some extent of people, although the latter movements may also be driven by complex historical, sociological, economic and cultural motivations (Chapter 3). Human–nature interactions are not limited to material interactions, but have also socio-cultural and psychological dimensions, such as living with the threat of wild animals (Marks, 1976). This chapter focuses on the material consequences of animal movements across the edges of protected areas. It looks at the impacts of such movements on the livelihoods of people living on these edges, whereas other chapters of the book focus on the causes of these movements (Chapter 7) and on the opportunities for the mitigation of their consequences (Chapter 9).

We first describe how the proximity of transfrontier parks (TFPs) poses additional constraints for the people, including physical aggression by wildlife, destruction of crops, predation of livestock and disease transmission. Simultaneously, living on the edge of such protected areas may provide benefits, such as additional grazing opportunities for livestock or revenues and game meat through community-based natural resource management (CBNRM) programmes.

Adopting a comparative and integrated perspective (including disease transmission as an integral component of human–wildlife conflicts resulting from animal movements), this chapter builds on recent case studies from Zimbabwe, South Africa and Mozambique. It seeks to identify key characteristics of these socio-ecological systems that are likely to reduce the magnitude of the negative consequences for people of animal movements across the edge of TFPs.

Different categories of human–wildlife conflicts

The implications of wild animals crossing the edge of protected areas and penetrating the neighbouring human dominated areas are manifold. The severity of these implications is a function of the density of people living adjacent to the protected area, the type of activities prevailing and the species of wild animal involved. Broadly, five categories of conflicts result from wild animals crossing the edges of TFPs (as illustrated in Figure 8.1): human death, human injury, crop raiding, predation of livestock and disease transmission. The next sections of this chapter evaluate how people living in TFCAs in Southern Africa are impacted by wildlife and how they perceive these constraints.

It is likely that only the most extreme cases of human–wildlife conflict are reported to authorities, as exemplified by a survey carried out in 2009 (Figure 8.1) in the Mbire District (Zambezi Valley, Zimbabwe; see Figure 3.5). Affected people

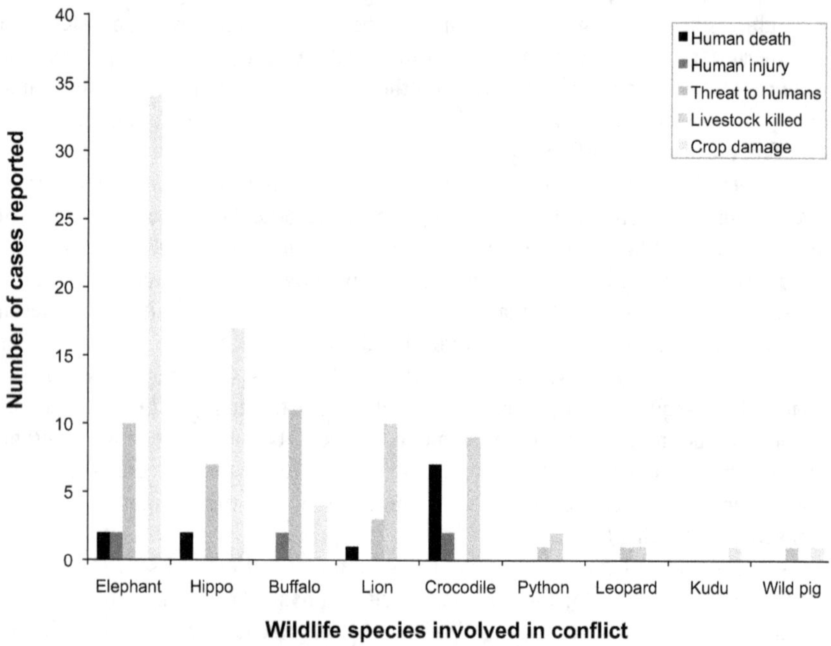

Figure 8.1 Numbers of human–wildlife conflicts and their sources in the Mbire District, Mid-Zambezi Valley, Zimbabwe. The bars indicate the number of cases recorded by Mbire Rural District Council during the period January 2006 to July 2009.

indicated that most crop damage is unreported, as the National Parks and Wildlife Management Authority (NPWMA) and the Rural District Council (RDC) are less responsive to cases of crop damage than during the previous decade. There is little sense in reporting crop damage, as this does not result in compensation. However, wildlife attacks on human and livestock are reported, despite a lack of action by either the NPWMA or the RDC, as they do not have the capacities to assist the people appropriately. In total, 14 human deaths due to wildlife were reported in Mbire District between January 2006 and July 2009, which were attributed to crocodile (57 per cent), elephant (14 per cent), hippo (14 per cent) and lion (7 per cent). During the same period, 6 persons were injured by crocodile, buffalo and elephant, and more than 35 cases of life-threatening situations (mainly by buffalo, elephant, hippo and lion) were reported. This is probably also an underestimate, as some cases go unreported to the RDC. In addition, the psychological impact of wildlife proximity should not be underestimated. The fear of dangerous wild animals can cause stress and prevent people from moving on foot, for instance children going to school or people moving at night for social gatherings.

Crop raiding

Subsistence farming is a major activity for people living in TFCAs in Southern Africa, with the possible exception of South Africa as will be discussed later. Destruction of crops by wildlife has long been recognized as a major constraint to the coexistence of man and wildlife, and a wide variety of vertebrates conflict with farming activities in Africa. These include birds, rodents, primates, antelope, buffalo, hippopotamus, bushpig and elephant (Lamarque *et al.*, 2009). While it is widely recognized that in most cases elephants and other large mammals do not inflict the most damage to subsistence agriculture, they are generally identified as the greatest threat to African farmers (Parker *et al.*, 2007). Birds, such as *Quelea*, and rodents can be responsible for greater crop losses than large mammals, as demonstrated by Marks (1976) in the Zambezi Valley. However, crop damage due to birds and small mammals occurs in many areas of Africa, regardless of the proximity of protected areas, and as such is not perceived as a direct consequence of animal movements across the edges. Using case studies from Zimbabwe, we evaluate how crop raiding by large mammals affects people living at the edge of TFPs.

In the Mbire District, a detailed survey by Baudron *et al.* (2011; see also Chapter 4 and Figure 3.5) provides insights on crop damage along a gradient where the lowest human population densities are found close to the protected area. Crop damage by wildlife decreases from all (100 per cent) heads of household interviewed in West Angwa (adjacent to protected area) reporting crop damage by wildlife compared with 58 per cent in Mushumbi Pools, which is further from the protected area (Table 8.1). The probability of crop damage has a significant impact on time invested in protecting fields. Whereas all of the interviewed farmers indicated that they guard their fields against wildlife in West Angwa, only 77 per cent did so in East Angwa and 55 per cent in Mushumbi Pools. Elephant and buffalo were the major species reported to cause crop destruction, as reported

Table 8.1 Frequency of crop raiding in the Mid-Zambezi Valley, Zimbabwe

Proportion of farms (%) affected by crop destruction by wildlife in three areas of increasing distance from the complex of protected areas Mana Pools NP–Sapi SA–Chewore SA (Mbire District, Mid-Zambezi Valley, Zimbabwe): West Angwa, East Angwa and Mushumbi Pools

	Geographic zone		
	West Angwa (n = 26)	*East Angwa (n = 47)*	*Mushumbi Pools (n = 99)*
Crops destroyed by wildlife	100	87	58
Crops destroyed by elephants	96	43	30
Crops destroyed by buffalo	73	23	3
Fields guarded	100	77	55

Source: Baudron *et al.* (2011).

Note*:* Interviews conducted during transect walks in 2008 (total sample $n = 172$).

earlier in the neighbouring district of Muzarabani (Parker and Osborn, 2001). Hoare (1999) also noted a reduction in elephant damage on crop land with distance from a protected area.

Intense human–wildlife conflicts occur in the KAZA TFCA in north-western Zimbabwe. The creation of Wankie Game Reserve in 1928 (now Hwange National Park; see Box 3.3) was followed by the supplementation of existing waterholes during the dry season in the northern part of the park. This increased surface water availability expanded the dry season habitat available to elephants from 35 to 75 per cent of the park area (Cumming, 1981). The elephant population of Hwange National Park (HNP), which is strongly linked to populations in Botswana, grew from approximately 2,000 elephants in 1928 to more than 10,000 when culling to reduce elephant impacts on woodlands started in 1966 (Cumming, 1981). The numbers culled during the 1960s and 1970s were based on underestimates of the numbers of elephants and made little impact on population growth. The introduction of improved census techniques in 1980 revealed many more elephants than expected (21,668 in 1983) and a population that was increasing at 5 per cent per annum. The DNPWLM decided to reduce the population to 12,000 elephants to maintain a density of about 0.75 elephants per km^2. Between 1984 and 1986, some 8,000 elephants were culled and the population was reduced to 13,000 (Cumming, personal communication). However, no further culls took place and the population rapidly expanded to reach *c.*35,000 in 1992 and a peak of 44,500 in 2001 (Dunham *et al.*, 2002). It appears that the elephant population in the park has stabilized around 35,000 individuals (Chamaillé-Jammes *et al.*, 2008), with a further 5,000 in the Matetsi complex and forest areas and less than 1,000 in surrounding communal lands (Dunham *et al.*, 2007). The high densities in HNP, which are among the highest in Africa, result in frequent dispersal of elephants into neighbouring farming areas and associated frequent conflict.

When questioned about the different wild herbivores commonly sighted in communal areas over the year, people from the Dete area (Figure 3.4) most often cited elephant, bushpig, baboon and kudu. People are mainly concerned about elephants and, to a lesser extent, bushpigs, which are known to raid maize fields during the planting and the ripening period. They suggest elephants were rarely seen in communal areas during the decade after the culling was stopped. Part of the area is bordered by a forestry concession interconnected with HNP. The border was well delimited by a veterinary fence until 2003, when it collapsed due to a combination of lack of maintenance and local community pressure to access grazing lands located inside the Sikumi Forest Area. Elephant damage is increasing, raising the concern of people and conflicts between the community and local authorities. Elephants are typically observed during the wet season (from November to June), but more than 70 per cent of the interviewees reported a peak activity of crop damage in March–April, when the grains are ripening. Selective pressure (Hoare, 1999) and food quality (Osborn, 2004) can influence behaviour and crop raiding by individual bull elephants. Other important factors that influence crop damage by elephants are human settlement density, the crops cultivated, farm infrastructure, and the farmers' efforts to guard the crops (Sitati *et al.*, 2003, 2005).

A small-scale crop damage survey (n = 30 fields) was carried out in 2009 in a community close to HNP, just before the harvesting time. The Magoli area is bordered by a forestry concession (Sikumi Forest) interconnected with HNP. Twelve of the 30 fields were significantly (more than 5 per cent of the crop stems destroyed) impacted by elephant and five by bushpig (Figure 8.2). In some cases, more than a third of the field was destroyed and, on average, elephant raids had a greater impact than other species (elephant: 22 per cent of the field destroyed; bushpigs: 11 per cent; livestock – i.e. cattle, goats and donkeys: 10 per cent). Maize and sorghum were raided more than millet (respectively 36, 22 and 14 per cent). Risk of crop damage decreased with distance from the forest area, and bull elephants were more often to blame than family herds. No elephant raids were recorded beyond the main dirt road in the communal land of Dete (for details, see Figure 8.2), probably due to a higher human activity. Bushpig damages were mostly observed in fields close to the savannah patches in low human density areas. The cumulated impact of livestock damage was greater than elephant, particularly along the roads which are often used by herders to access grazing areas.

People grow crops essentially for subsistence, although they can trade their production surplus to get basic necessities. Silewu ward is classified as Natural Region V (Vincent and Thomas, 1960), characterized by low and erratic rainfall and considered unsuitable for rain-fed agriculture. Damage inflicted on crops by wildlife and livestock have direct impacts on food security and, more broadly, on people's welfare. As an example, the estimated sorghum losses on the different impacted fields ranked between 5 and 150 kg of grain. People often brew and sell sorghum beer (US$1 for 2 litres) to secure incomes for school fees (on average US$1 per child per day). Ten percent of the investigated households mentioned that they left their previous homestead for lands located further from HNP because of conflicts with elephants and the resulting difficulty of producing sufficient food

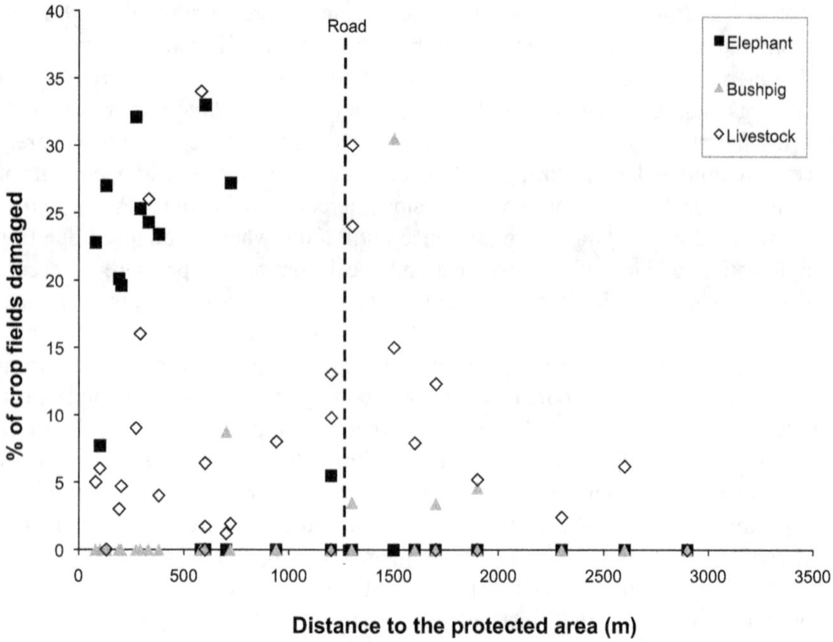

Figure 8.2 Crop damage by large herbivores in Magoli area, northern edge of Hwange National Park, Zimbabwe. Percentage of destruction in each monitored field (n = 30 fields, May 2009) in relation to the distance from forestry land.

from their farming activities. Another 5 per cent abandoned part of their cropping land for the same reasons. Due to the economic crisis prevailing in Zimbabwe for the past decade, the CAMPFIRE programme in this area has collapsed. People are now waiting for NPWMA – often cited as the owner of wildlife (92 per cent) – to mobilize and help them to coexist with wildlife. Less than 10 per cent of the interviewed households mentioned benefiting from wages and 52 per cent suggested that employment would reduce poaching pressure.

Livestock predation by wildlife

Killing of livestock by wild predators is a problem for most communities living at the edge of TFPs in Southern Africa. Data from a study of eight villages (Mhinga Traditional Authority, Vhembe District, South Africa; see Figure 8.5) falling within a 15-km radius of the north-western border fence of Kruger National Park (KNP) illustrates the magnitude of the problem.[1]

Although only 11 per cent of the households in the area own cattle, cattle keeping in this area has important socio-economic and cultural functions. Both cattle-owning and non-cattle-owning households indicated that they derived direct and indirect benefits from cattle farming in the area. Of the households without cattle, 55 per cent were interested in cattle farming, and 13 per cent cited problems with wildlife as the main reason why they would not want to own cattle. Some of the households without cattle at the time of the survey had owned cattle and lost them to wild predators in previous years.

Some 25 per cent of the 540 sampled households had cattle killed by wildlife between June 2005 and June 2008. More households in villages close to the park reported livestock losses (32 per cent) than those further from the park (12 per cent). The most common cattle predators reported were lion, leopard, hyena and crocodile. Most (64 per cent) of the cattle were reported to have been killed in the 'veld', although some (36 per cent) were killed in the kraal. Cattle were killed by lions mainly at night (60 per cent). Farmers explained that in some cases cattle were lost and spent the night outside the kraal, which exposed them to a greater risk of being killed by wild predators. In Matiyani village, which experienced the greatest loss, cattle had to walk long distances to a water point. This increased the risk of attack as those that lagged behind were out of sight of the herdsmen.

Loss of cattle affects the household's direct and indirect benefits from livestock. Direct benefits derived include draught power (27 per cent of households), access to manure for fertilizing crop fields (43 per cent), milk (41 per cent) and meat. Gross financial losses to farmers from depredation were estimated at US$106 per

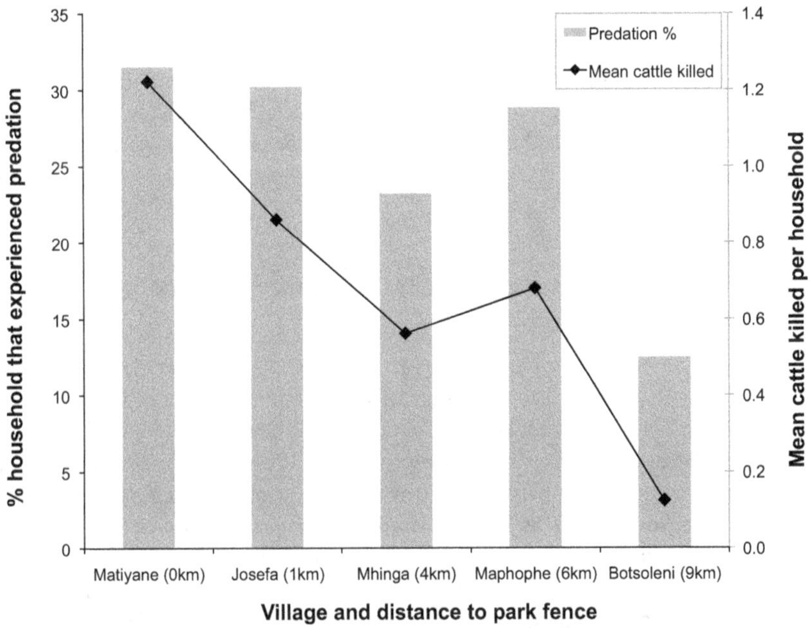

Figure 8.3 Cattle predation by wildlife in villages situated at various distances from Kruger National Park. Survey done in July 2008, 270 households interviewed in Mhinga Traditional Area, South Africa. The bars indicate the % of households in each village experiencing cattle predation in the period (July 2005 to June 2008), and the line indicates the mean number of cattle per household killed during the same period.

Source: Chaminuka *et al.* (in press).

annum per household in the period 2005–2008, which was nearly 11 per cent of the average annual household incomes of US$960 in this area (Pauw, 2005). The loss induced by wild predators represented 0.7 cattle/household in 3 years, a considerable loss in an area where the average herd size is nine, and about 42 per cent of the households had less than five cattle. Some farmers had made efforts to secure their cattle kraals by putting fencing at the base to prevent predators from crawling in.

The species of wild predators most frequently involved in livestock attacks varied slightly between sites located at the periphery of the protected areas. Species responsible for the majority of livestock losses were lion, hyena and leopard in the north-western periphery of KNP, crocodile and lion in Mbire District (Figure 8.1), and hyena and lion in the communal lands adjacent to HNP (Chapter 7). The impact of livestock predation on people's livelihoods was substantial in all cases. Communities living in Vhembe District had negative perceptions about KNP due to wildlife predation on livestock, supporting earlier findings by Anthony (2006). In addition, there were no uniform procedures for reporting predation. Most (73 per cent) of the households that suffered losses of cattle to wild predators indicated they had reported this loss to either the local traditional authority, government officials, the Hlanganani Forum, a civic society organization, or to KNP officials. In none of the cases had farmers received compensation for cattle lost between June 2005 and June 2008, which increased their resentment to the KNP. In group discussions to explore solutions to the problems created by wild predators, farmers suggested that they should be allowed to kill the predators outside the park, or they should be employed by KNP to patrol and repair damaged fences, and that patrols of the fences by KNP should be increased.

Diseases as a consequence of livestock and wildlife crossing the edges

Disease transmission is arguably one of the major obstacles to the coexistence of wildlife and livestock in sub-Saharan Africa (Bourn and Blench, 1999). Movements of both humans and animals occur across the boundaries of TFPs in Southern Africa, often driven by resource gradients (Chapter 7), which results in direct and indirect contact between wild and domestic animals, potentially leading to disease transmission. This is considered a major problem by people living in areas where wildlife and livestock frequently come into contact. In 2006–2007 in the Southeast Lowveld of Zimbabwe,[2] farmers ($n = 431$) mentioned three main reasons why wildlife represented a threat to them: (1) crop destruction (90.6 per cent of farmers questioned); (2) predation of livestock (90.4 per cent); and (3) diseases spread to cattle (90.7 per cent). The frequency and intensity of interactions with wildlife is largely dependent on the existence of effective fences. In contrast to the responses in the Southeast Lowveld, where the fenced boundaries are highly porous, livestock owners from Mhinga Traditional Authority (Vhembe District, South Africa) living across the fence of KNP perceived livestock predation to be a much greater problem (89 per cent) than disease transmission (3 per cent) (Chaminuka, unpublished data, 2008; $n = 270$ farmers interviewed). This suggests that the hard-edge of KNP reduces the contacts between livestock and the wild reservoir of diseases, but the fences are not as efficient in preventing predators from raiding livestock outside the park.

Movements of livestock and wildlife across the edges of protected areas are an avenue for the transmission of pathogens in either direction: from wildlife to livestock and vice-versa. In some cases, the direction of pathogen transmission has been documented using molecular tools in Southern Africa (e.g. rabies, Sabeta *et al.*, 2003; Foot and Mouth Disease, Vosloo *et al.*, 2005; bovine tuberculosis, Michel *et al.*, 2006). The role as disease reservoir played by wildlife species is well established, for example buffalo act as a reservoir of *Theileria parva lawrenci* responsible for corridor disease. But in most of the cases a role is suspected but not sufficiently documented, as the occurrence of the same disease is no proof of the transmission of pathogens between wildlife and livestock. Diseases occurring in livestock and wildlife at the edge can be categorized into: (a) wildlife-maintained diseases, to which wildlife are resistant or silent carriers of infection and (b) multi-species diseases, which have serious outcomes in both livestock and wildlife. Important wildlife-maintained diseases are many, including Foot and Mouth Disease (FMD), malignant catarrhal fever, trypanosomiasis, theileriosis (corridor disease), ehrlichiosis (heartwater), lumpy skin disease and Rift Valley fever (Bengis *et al.*, 2002). Multi-species diseases, which include anthrax, rabies, brucellosis and bovine tuberculosis (BTB), occur in both wildlife and domestic livestock, and transmission can occur in both directions. Uniquely, these diseases generally have a fatal outcome in both wildlife and domestic livestock, and are frequently zoonotic, meaning that they can be transmitted to humans (Bengis *et al.*, 2002).

Endemic in Southern Africa, rabies is responsible for numerous human deaths (Knobel *et al.*, 2005), and also represents a major threat for the conservation of highly threatened African species such as wild dog (*Lycaon pictus*). BTB, which is discussed further below, is a disease that affects wildlife in Southern Africa (Renwick *et al.*, 2007) and was originally introduced through livestock. Although livestock diseases that infect wildlife can have a major impact on conservation, we focus our attention on livestock diseases that affect the people living at the periphery of TFPs in Southern Africa.

Frequency of wildlife-related diseases in livestock populations across fenced/unfenced interfaces

Fencing to prevent animal movements and contacts has long been advocated for disease control at the wildlife–livestock interface (see review by Ferguson and Hanks, 2010). The attitude of local populations towards game fences varies according to the socio-cultural and historical context, ranging from an 'institutionalized respect for fences' in Botswana, to clear opposition, as illustrated in Zimbabwe during the past decade by the destruction of the FMD fence around

Gonarezhou National Park (GNP) and other protected areas (Dube *et al.*, 2010). Here we focus on the frequency of wildlife-related diseases in livestock populations adjacent to the unfenced boundary of HNP (Zimbabwe; Figure 3.4), as compared with the fenced boundary of KNP (South Africa; Figure 3.2).

HNP hosts a large variety of wild herbivores and carnivores which are known to play an important role in the epidemiology of diseases that are transmissible between wildlife and livestock. A retrospective study covering a period of 13 years in communal areas surrounding HNP (January 1995 to July 2007; Marange and Marimwe, 2008) collected data on diseases transmissible between wild and domestic animals.[3] Wildlife-maintained and multi-species diseases contributed over 70 per cent of the total disease cases reported ($n = 2989$). The study detected an increased occurrence of wildlife-maintained and multi-species diseases between 1995–2000 and 2001–2006 (Figure 8.4a). The increase started in 1999–2000, just after the economic downturn in Zimbabwe, which began in 1997, but before the resettlement of communal farmers in former game ranches, which increased interactions between wildlife and livestock. In addition, a significantly higher number of cases of these diseases were reported from dip tanks located within a radius of 20 km from the national park boundary compared with those further from the park (Figure 8.4b).

During the period 1995–2009, a number of outbreaks of wildlife-related diseases were reported in livestock populations living in Mpumalanga Province at the periphery of the fence with KNP (Du Plessis, 2004, 2005; see Figure 8.5). The most significant included two outbreaks of FMD, 15 outbreaks of corridor disease and one outbreak of anthrax. All of these events were believed to be linked directly with KNP, and posed serious risk for public and veterinary health although the veterinary services controlled the outbreaks effectively. Brucellosis is another zoonosis affecting wildlife, livestock and humans, which occurred at a prevalence of 0.7 and 1.7 per cent in cattle herds tested between 1999 and 2004 in communal and commercial farms in Mpumalanga Province. Since the 1990s, brucellosis has been controlled through vaccination of heifers. While it is possible that the disease spread from wildlife to cattle, as several wild animal species tested serologically positive in the KNP (Chaparro *et al.*, 1990; Godfroid, 2002), this has not been confirmed.

BTB spread from cattle to wildlife more than 40 years ago in the south of KNP (Michel *et al.*, 2006; see also Figure 8.5). Between 1995 and 2003, no cases of BTB were detected in cattle surrounding KNP, but the disease was detected in a cattle herd in 2003 south of the KNP (Du Plessis, 2004), seemingly originating from infection in wildlife. This cattle herd had contracted FMD and suspected corridor disease infections following contact with stray buffalo after floods in 2000 (Jori *et al.*, 2009). In 2004, tuberculosis was diagnosed in two additional nearby cattle herds (Du Plessis, 2005). In June 2009, a new outbreak occurred in communal lands (J. Kotze, personal communication). Therefore, although BTB appeared to have been successfully confined to wildlife, the disease has occasionally been transmitted back to cattle, possibly triggered by unpredictable events leading to contact between buffalo and cattle, such as floods or droughts.

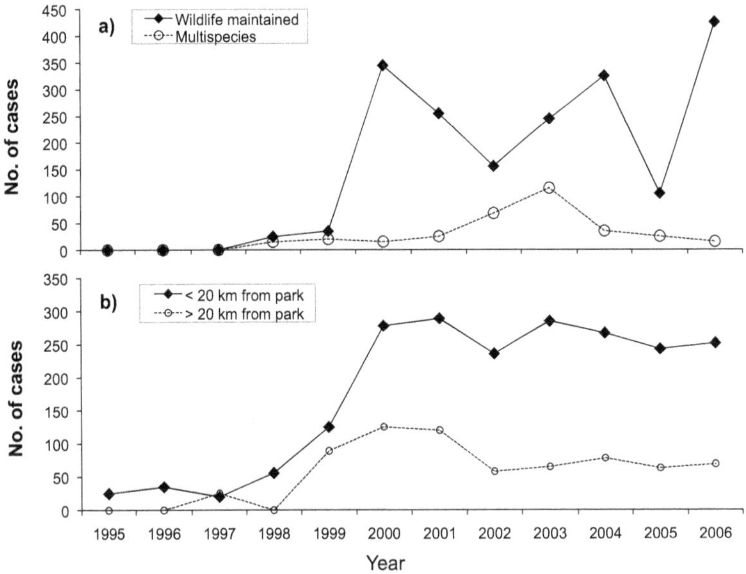

Figure 8.4 Annual occurrence of wildlife-maintained and multi-species diseases in livestock, during the period 1995–2006, Hwange District, Zimbabwe. Number of cases reported according to (a) year and (b) distance from Hwange National Park. Figure adapted from Marange and Marimwe, 2008.

Fences at the edge of TFPs are clearly effective in reducing both direct and indirect contacts between wildlife and domestic livestock, and the resulting disease transmission. Apart from the fences, additional factors contributed to the differences observed between HNP in Zimbabwe and KNP in South Africa. In addition to the destruction of the game fence, increased transmission of diseases was favoured by difficulties faced by the Zimbabwean veterinary services since the early 2000s to implement efficient disease surveillance and control, and increased human densities at the periphery of the HNP with inadequate grazing and watering resources particularly during the dry season. It could also be linked to the transformation of former game ranches into small-scale subsistence farming land during the agrarian reform period and the illegal livestock movements that followed in some areas of Zimbabwe. In South Africa, diseases in livestock and wildlife are regularly monitored within and outside KNP, and the veterinary services benefit from adequate financial, human and technical resources to implement an efficient disease surveillance and control. Despite the existence of a fence, contacts between wildlife and livestock still occur both inside and outside KNP (see Chapter 7). Most wildlife-related diseases occurring in livestock populations living at the edge of unfenced protected areas (e.g. Malipati Communal Land and Hwange District, Zimbabwe) also affect livestock populations living on the fenced edge of KNP (e.g. FMD, brucellosis, theileriosis and anthrax), albeit with less frequency.

The social, economic and health impacts of wildlife-borne diseases on people living at the periphery of KNP are less severe than in Zimbabwe for several reasons. First, rural households in South Africa rely less on agriculture for their livelihoods in comparison to the rest of the region. Second, although the game fence around KNP does not prevent all contact between wildlife and livestock, it reduces the frequency of contacts. Third, the veterinary services implement livestock disease surveillance, prevention and control in the buffer zone adjacent to the park. The comparison of livestock disease incidence at the periphery of KNP with HNP and GNP in Zimbabwe emphasizes the need for fence maintenance and specific, and often costly, veterinary programmes to mitigate the burden that wildlife-borne diseases represent for people living in TFCAs in Southern Africa.

Spread of diseases by wildlife across international boundaries of TFCAs

One of the main objectives of TFCAs is to promote, or restore, for conservation purposes, the movements of wildlife between protected areas situated across international boundaries (Bengis, 2005). The longer term plans for the GLTFCA focus on the development of wildlife-based tourism with freedom of movement for wildlife and tourists across international borders. Thus interactions between wildlife, livestock and humans living in the conservation area can be expected to increase significantly (Michel *et al.*, 2006). In this section, we describe the distribution of outbreaks of two diseases (i.e. FMD and BTB), which have been demonstrated to be spread by wildlife across international boundaries of the GLTFCA, and analyse the consequences of these diseases for the people living directly adjacent to the protected areas where the diseases occur.

In Zimbabwe, a total of 591 primary and secondary outbreaks of FMD were recorded between 1931 and 2006, with the majority (70.1 per cent) occurring in the endemic FMD vaccination and non-vaccinated control/buffer zones of the Southeast Lowveld and the north-west (Condy, 1979; Zihumo and Mapuvire, 2006; Caron *et al.*, 2010). Zimbabwe was divided into four zones for FMD control: a wildlife zone; a vaccination zone adjacent to the wildlife zone ('red' zone) in which all cattle were vaccinated every six months; a non-vaccinated buffer zone ('green' zone) from which all susceptible animals were tested serologically negative before they were moved to; the FMD-free 'clear' zone (Anderson *et al.*, 1993). Between 1931 and 2000, the majority of FMD outbreaks (96.5 per cent) were confined to the control/buffer zones surrounding the national parks. However, between 2001 and 2006, the disease spread into the 'clear' zones from which cattle were exported. The majority (57.7 per cent) of the outbreaks occurred on properties within a radius of 60 km from national parks, but a few occurred as far as 140 km away (Caron *et al.*, 2010). Seven primary outbreaks of the disease during this period involved contact between cattle and wildlife. FMD outbreaks have had serious adverse effects on the meat industry in Zimbabwe. Before the FMD outbreaks, Zimbabwe used to export a beef quota of 9,100 tons to the EU and 5,000 tons to South Africa, which has now been stopped completely. Although this caused major economic loss in Zimbabwe, people living on the edge were not directly affected by the export ban, as livestock and livestock products originating from FMD-infected areas at the edge of FTPs cannot be legally and freely marketed outside the area.

Buffalo in the Limpopo National Park (LNP) (Mozambique) occasionally come into contact with cattle (e.g. Shingwedzi River basin, south-western corner of the LNP; see Figure 8.5). Preventive measures against FMD in Mozambique encompass movement control, vaccination, surveillance and monitoring along border regions, and outbreak control, although their efficiency is limited due to insufficient human and financial resources. Routine vaccination is only implemented in infected and buffer zones and those areas at higher risk (Maputo, Gaza, Beira corridor, Manica and Sofala). According to official records, the first cases of FMD in Mozambique were reported in 1937. Between 1986 and 2001, no FMD outbreak was reported, but 23 outbreaks were reported in southern Mozambique in 2002–2003, between the GNP in Zimbabwe and the LNP in Mozambique. These outbreaks were an expansion of the outbreaks that were occurring in Zimbabwe at that time and the spread was due to several factors, most likely linked to theft and illegal transboundary movements of cattle (see Chapter 6).

In South Africa, FMD normally only occurs in African buffalo in KNP and associated private conservation areas, the virus being maintained by buffalo (Thomson and Bastos, 2004). While the last outbreak of FMD in livestock in the KNP buffer zone occurred in 1983, seven outbreaks – with confirmed epidemiological connection to the KNP – occurred between 2000 and 2009 (Vosloo *et al.*, 2002a, 2002b; Reininghaus, personal communication). Some of these outbreaks were near the perimeter of the free zone, threatening agricultural exports on one occasion (Jori *et al.*, 2009).

BTB was introduced to Southern Africa during the nineteenth century through cattle imported from Europe (Huchzermeyer *et al.*, 1994). It is now emerging as an important disease in Southern African wildlife, with far reaching health implications for wildlife and for livestock and humans (Michel *et al.*, 2006). Although the first cases of BTB infection in African buffalo from KNP were diagnosed in 1990, it is established that cattle were responsible for the introduction of the disease in the park (Bengis *et al.*, 1996) probably during the late 1950s. Since then, more than 13 wild species in KNP have been confirmed to be infected by BTB (Michel *et al.*, 2006), and the disease has spread northwards in buffalo populations (Figure 8.5). Some 50 years after its presumed introduction from cattle in the south-eastern corner of the park, BTB was diagnosed in a buffalo only 6 km south of the Limpopo River (Keet, unpublished data, 2005). By 2008, BTB had spread to buffalo in the southern part of GNP in Zimbabwe (Mabalauta; see Figure 8.5), and molecular typing confirmed that it originated from KNP (de Garine-Wichatitsky *et al.*, 2010).

Once established in a native free-ranging maintenance host, it is unlikely that BTB can be eradicated (De Lisle *et al.*, 2002, Michel *et al.*, 2006). South African

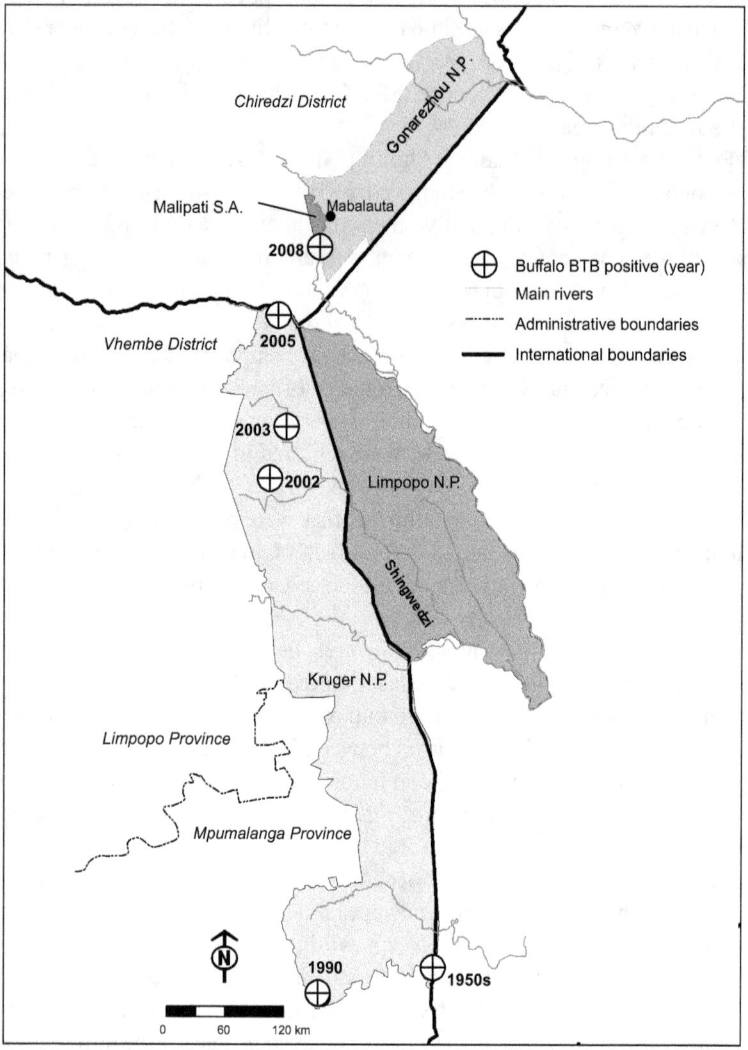

Figure 8.5 Spread of bovine tuberculosis by wildlife across boundaries of the GLTFP. Approximate locations (crosses) and dates of confirmed BTB cases in African buffalo. Map redrawn using several sources of information.[4]

national parks have not been able to control the spread of the disease in KNP, despite significant efforts to monitor and experimental development of control methods (e.g. vaccination). Despite the presence of the western KNP boundary fence, it is suspected that BTB infection spilled back to livestock populations (Du Plessis, 2004). But as wildlife–livestock–human interactions are more intense in and along the edges of LNP and GNP,[5] as described above (see also

Chapter 7), the disease could have significant future consequences for human and livestock health in the GLTFCA (Michel *et al.*, 2009).

People's perception of wildlife-borne diseases in livestock

Perceptions of wildlife-borne diseases among cattle keepers in the Southeast Lowveld of Zimbabwe (Table 8.2) revealed a large discrepancy between the diseases mentioned by farmers and those reported by the Department of Veterinary Services (DVS) technicians in the same area.[6] FMD was the most commonly mentioned disease by both, although it was mentioned by only 45 per cent of the farmers, whereas it represented more than 98 per cent of the DVS records. The frequent mention of FMD probably does not reflect the real impacts of the disease on livestock production in the Malipati area. It is more likely the result of the strong emphasis, and associated communication to the public, of DVS during the past decades to control FMD by means of vaccination and fencing. The difficulties faced by Zimbabwean government services during the economic crisis resulted in cessation of the FMD vaccination programme in the Malipati area and lack of maintenance of the FMD fence. Other diseases mentioned by livestock owners most likely had stronger impacts than FMD. These include tick-borne diseases (anaplamosis, ehrlichiosis and babesiosis), which increased after dipping was disrupted and have devastating effects on livestock numbers in such circumstances,[7] as well as lumpy skin disease and blackleg, which used to be controlled by vaccination. For most of the diseases mentioned, wildlife species have been suspected or demonstrated to transmit the pathogen to livestock (Table 8.2). Livestock keepers in Malipati were generally aware of the possible role played by wildlife, and especially of the role of reservoir of FMD virus played by the African buffalo.

Several diseases were not mentioned by either livestock keepers or the DVS although they could represent serious threats for public health. Brucellosis and Rift Valley fever were not mentioned, although a serological survey indicated prevalences of 8.5 and 19.8 per cent, respectively in livestock (PARSEL projects, unpublished results; Gomo, 2010). Anthrax was mentioned by only 0.3 per cent of the farmers interviewed, although it is an important zoonosis responsible for significant mortality of wildlife, livestock and sometimes human beings. Rabies was also not mentioned, probably because of the focus of questions on cattle farming. Lastly, although not a zoonotic disease, no mention was made of corridor disease, another tick-borne disease carried by the African buffalo, which has been diagnosed in livestock from the Malipati area.

In the context of an intense wildlife–livestock interface and limited disease surveillance and control by the veterinary services, farmers in Malipati perceived disease transmission as a major threat to livestock production. They generally had an accurate perception of the most important diseases affecting their cattle, and of the potential epidemiological role played by wildlife. However, although the livestock keepers recognized the disease risks associated with wildlife for their cattle, they were apparently not aware of sanitary risk for themselves. Zoonotic risks were never mentioned in the discussions. In the absence of reliable data,

Table 8.2 Livestock diseases in the Malipati area, Southeast Lowveld, Zimbabwe: disease and causative agent, frequency of citation by farmers

Disease and causative agent	*Cited by farmers (%)*	*Recorded by DVS (%)*	*Serological confirmation in livestock*	*Transmission from wildlife to livestock (Ref)*
Foot and Mouth Disease (*aphtovirus*)	44.2	98.2	Yes*	Yes (1, 3)
Lumpy skin disease (*capripoxvirus*)	20.8	0.2	Yes	Uncertain (1, 5)
Blackleg (*Clostridium chauvoei*)	19.2	0.5	*n.t.*	Unknown
Anaplasmosis (*Anaplasma marginale*)	7.4	0.4	Yes	Suspected (12)
Ehrlichiosis (*Ehrlichia ruminantium*)	5.7	0.5	Yes	Yes (1)
Bovine ephemeral fever (*ephemerovirus*)	1.9	—	*n.t.*	Suspected (10)
Babesiosis (*Babesia bigemina*)	0.5	0.2	Yes	Uncertain
Anthrax (*Bacillus anthracis*)	0.3	—	*n.t.*	Yes (1, 4)
Brucellosis (*Brucella abortus*)	—	—	Yes	Yes (1, 2, 11)
Rift Valley fever (*phlebovirus*)	—	—	Yes	Uncertain (1, 6, 8, 9, 10)
Theileriosis (*Theileria parva*)	—	—	Yes	Yes (1, 7)
Total	$n = 823$ diseases cited	$n = 2{,}091$ records	$n > 400$ cattle sampled	

Legend: '—' not mentioned ; '*n.t.*' disease not tested; * FMD serological tests used did not differentiate vaccination from infection.

References cited (see full details in reference list): (1) Bengis *et al.*, 2002; (2) Godfroid, 2002; (3) Thomson *et al.*, 2003; (4) Hugh-Jones and de Vos, 2002; (5) Davies, 1982; (6) Davies, 1975; (7) Grootenhuis, 1988, (8) Evans *et al.*, 2008, (9) Paweska *et al.*, 2008, (10) Anderson and Rowe, 1998, (11) Madsen and Anderson, 1995, (12) Norval *et al.*, 1984.

Notes: Percentage of responses to questionnaire survey in 2006–2007: $n = 431$ livestock owners, multiple responses allowed; proportions of all disease cases reported by Department of Veterinary Services confirmed presence of disease in livestock (serological tests carried out on livestock samples from Malipati area in 2007–2008), and confirmed transmission from wildlife to livestock according to the literature.

it is difficult to assess whether zoonotic diseases, such as BTB or brucellosis, represent a public health problem in the Malipati area, especially compared to other prevalent diseases such as malaria or HIV/AIDS. The results suggest that a comprehensive education programme on disease risks at the wildlife–livestock–human interface is needed.

Benefits of livestock and wildlife crossing the edge

Living on the edge also provides opportunities for the people to directly benefit from the proximity of wildlife and other goods and services provided by protected areas. These opportunities are explored in Chapter 9, and we focus here on direct benefits that people living on the edges may receive from wild and domestic animal movements across the boundaries of protected areas.

Roaming of wild animals in communal lands adjacent to protected areas represents the greatest opportunity for communities to benefit directly through CBNRM programmes. The proximity of abundant wildlife offers access to highly prized game meat, often through illegal channels ('poaching', see Chapter 7) or legally when safari operators or culling operations provide meat of animals killed in the communal land (e.g. impala meat in Nyaminyami District, Zimbabwe; de Garine and de Garine-Wichatitsky, 1999), or killed inside the national park during culling operations (e.g. in GNP during the 1992 drought). Although the success of CAMPFIRE has declined, some material and financial benefits have been provided to people living on the edges of protected areas in Zimbabwe. Between 1989 and 2006, CAMPFIRE income totalled nearly US$30 million, mostly from safari hunting directly linked to the presence of wildlife in communal lands neighbouring protected areas (Taylor, 2009). Some 52 per cent of this income was allocated to wards and villages for community projects and household benefits. The utilization of these funds varied greatly between districts. The amount of money generated per household was generally small (less than US$10 per household during the period 1989–2001). However, CAMPFIRE dividends contributed to the rehabilitation of community infrastructure such as grinding mills or schools, and were also used to purchase grain for drought relief in some districts (Rihoy *et al.*, 2007; Taylor and Murphree, 2007).

As we have seen in Chapter 7, cattle owners benefit from the proximity of protected areas, which provide access to grazing and water resources, especially during dry periods. Sending cattle to graze in national parks, forest or safari areas is a risky strategy, not only because of the risk of predation or disease, but also because severe fines and even confiscation of the cattle may result. In South Africa, farmers living next to the KNP received subsidized fodder for cattle in dry years, free mandatory weekly dipping and free cattle vaccinations for FMD, black quarter and anthrax from the Department of Agriculture. Farmers living next to GNP in Zimbabwe used to benefit from similar compensation measures (free vaccination against FMD, anthrax and rabies) but these services have been disrupted during the past decade.

Conclusion

The case studies from Zimbabwe, South Africa and Mozambique described in this chapter illustrate that coexisting with wildlife is often costly for people living on the edge of TFPs in Southern Africa. Local communities currently living in the GLTFCA, KAZA TFCA and Mana Pools-Lower Zambezi TFCA consistently suffer from the proximity of wild animals, including problems of physical threat, crop raiding by large mammals, loss of livestock to predators and transmission of diseases to livestock. People living closer to the edges of protected areas suffer substantially more than those living further away, although some highly contagious wildlife-borne diseases can reach livestock populations located far from the boundaries of TFCAs. Further, wild animals are classified under Roman-Dutch law as *res nullius*, i.e. being owned by nobody until captured or killed; although governments claim ownership on behalf of the state (Muir-Leresche and Nelson, 2001; Scholes *et al.*, 2007). This implies that farmers cannot claim for crop damages and livestock loss, and they are not compensated. At the same time, people suffering from attacks are not allowed, by law, to kill the wildlife responsible. Unsurprisingly this is perceived to be unfair and creates strong resentment as access to game meat would represent a form of compensation, although it would benefit the whole community rather than the affected individuals.

Several factors can attenuate the impacts of wildlife on the livelihoods of people living on the edges in Southern Africa. In South Africa, subsistence farming has been virtually eliminated, and nearly two-thirds of rural South African households earn nothing at all from agriculture (Seekings, 2000). Although livestock keepers living on the edge of KNP can suffer significant losses, the livelihood impacts of crop raiding by wildlife in South Africa are less than in countries where agriculture plays a much more central role in household livelihoods (Twine and Magome, 2008). The presence of fences also reduces the frequency and magnitude of problems associated with wildlife movements across the edges, as illustrated by the less frequent predation and incidence of wildlife-borne diseases in livestock at the periphery of the well-maintained fenced interface with KNP, as compared with the unfenced HNP and GNP. There are numerous social, economic and conservation drawbacks to reliance on fences (Taylor and Martin, 1987; Lamarque *et al.*, 2009; Ferguson and Hanks, 2010), but they are an effective tool in controlling animals' movements, especially if maintained properly (O'Connell-Rodwell *et al.*, 2000). However, no fence is fail-safe in preventing wildlife from making forays into communal lands and, regardless of the frequency or severity of the impacts, incidents of crop raiding by wildlife impact negatively on local attitudes towards protected areas. The fact that people are seldom compensated for losses incurred by wildlife is a major source of dissatisfaction (Spenceley, 2005; Anthony, 2007) in all three countries considered in this chapter.

To some degree, the current absence of efficient management of human–wildlife conflicts due to animal movements across the edges of TFCAs is a result of 'social-ecological scale mismatch' (Cumming *et al.*, 2006). The constraints faced by local farmers are the results of ecological and epidemiological processes occurring at local (local movements across conservation edges as a result of resource gradients) or regional scales (migrations or regional movements of wildlife, induced or constrained by TFCAs), whereas policies and control programmes are dictated at national or international level. Community-based animal health systems for instance, have been advocated as a way to improve livelihoods of remote, marginalized and underserved livestock keeping communities, through improved productivity and access to markets (Grahn and Leyland, 2005). The current schemes implemented regarding livestock health, including vaccinations, acaricide treatments or other prophylactic actions, do not address the local priorities as perceived by the livestock keepers themselves. This is largely because livestock disease management is driven by government veterinary services at national level, which often concentrate their limited resources on the control of transboundary animal diseases (OIE, 2003) to allow international trade, which is regulated by policies negotiated at the international level. The perception of livestock keepers regarding locally important diseases, and the potential role played by wildlife species, is rarely appreciated or taken into account when control measures are implemented. One of the main limitations of community-based animal health systems is the lack of voice of marginalized communities in policy processes (Grahn and Leyland, 2005).

It is anticipated that the creation of the GLTFCA, for instance, will increase significantly the movements of wildlife and tourists across international borders, ultimately resulting in increased interactions between wildlife, livestock and humans living in the conservation area (Michel *et al.*, 2006). The need for an integrated, interdisciplinary approach on animal health issues, including BTB, has been identified in a framework document by the AHEAD Working Groups (Osofsky *et al.*, 2005). The Peace Parks Foundation developed veterinary research objectives to address the concern raised by BTB and other diseases at the interface. Similarly, feasibility studies of the KAZA TFCA have identified the mitigation of human–wildlife conflicts (especially in 'hotspot' areas of conflicts with elephants) as a major priority for interventions (Metcalfe and Kepe, 2008).

However, mitigation of the costs of human–wildlife conflicts linked to animal movements, which should emphasize locally based solutions rather than top-down solutions, is only part of the equation for the acceptance of TFCAs by communities living at the edge. Another option is the improvement of their livelihoods through direct benefits that exceed the costs of living with wildlife (Metcalfe and Kepe, 2008; Chapter 9). Supporters of TFCAs have repeatedly claimed that communities will ultimately benefit as TFCAs will bring employment through the development of eco-tourism, a dream that is increasingly contested. However, little has been done to alleviate the additional burden represented by the proximity of wildlife, or to develop initiatives generating direct benefits from wildlife-based activities for communities living on the edge. Reforms in tenure and resource access rights, and the development of adaptive co-management arrangements between protected areas and neighbouring areas and communities are crucial for the sustainability of TFCAs (Cumming, 2008). Managing the movements of wild and domestic animals across the boundaries of TFPs to balance costs and benefits for communities (predation on crop and livestock, disease transmission, access to water and forage resources . . .) while meeting conservation goals presents several challenges. But there are opportunities for technical and institutional innovations and experimentations (see Chapter 9), such as the commodity-based trade of livestock products from TAD-infected areas (Thomson *et al.*, 2004; Thomson, 2008), negotiated controlled grazing by livestock inside designated areas in protected areas, or improved kraal designs and cropping practices in order to reduce wildlife predation (Lamarque *et al.*, 2009). Such initiatives should be promoted and implemented on the ground, as people living on the edges of TFPs expect more than a vague promise that eco-tourism will ultimately solve their problems associated with the proximity of wildlife.

Acknowledgements

The authors are grateful to Russel Taylor and Steve Osofsky for comments on an earlier version of this chapter. Original material and personal communications were kindly provided by Frédéric Baudron, Chaka Chirozva, Eléonore Hellard, Dewald Keet, J. Kotze, Steven Matema and Bjorn Reininghaus. We also thank Godfrey Pachavo for assistance in drawing Figure 8.5. We are grateful to the Department of Veterinary Services and the National Parks and Wildlife Management Authority of Zimbabwe, whose staff have actively participated in the field surveys and provided useful information. Finally, we would like to express our sincere gratitude to the local communities and traditional leaders of the areas where the various surveys referred to in this chapter have been conducted, with a very special thanks to the communities in Chiredzi District, Mbire District and Hwange District in Zimbabwe, and in Vhembe District in South Africa.

Notes

1 Data on the number of livestock killed by wild predators was collected through a household survey, inspection of veterinary records, farmer dip registers and key informant interviews. The household survey covered a total of 270 cattle-owning households, and 270 non-cattle owning households selected through stratified random sampling. Focus group discussions with cattle keepers followed up on issues raised in the survey.

2 Ten dip tanks attached to the Malipati Animal Health Centre (Chiredzi District, Zimbabwe) were included in the survey: Malipati, Dumisa, Muhlekwani, Pahlela, Chishinya, Maose, Rutandare, Samu, Mukwe and Chibhavahlengwe. An individual questionnaire survey was administered with 431 livestock keepers attending the dip tanks.

3 A total of 28 disease conditions (as reported by field staff of the Department of Veterinary Services, Hwange District, without laboratory confirmation) were recorded during 1995–2007 in cattle in Hwange District, with wildlife-maintained (8) and multi-species (7) diseases together contributing over 53 per cent of the disease conditions reported. The most frequent diseases included lumpy skin disease (44.9 per cent), dermatophilosis (21.4 per cent), ehrlichiosis (14.6 per cent), FMD (9.1 per cent), theileriosis (4.1 per cent), anaplasmosis (2.6 per cent), brucellosis (1.3 per cent), rabies (0.2 per cent), malignant catarrhal fever (0.2 per cent), Rift Valley fever (0.2 per cent) and anthrax (0.1 per cent).

4 Map redrawn using information from Bengis *et al.* (1996), Kloeck (1998; cited by Michel *et al.*, 2006), Grobler *et al.* (2002), Keet (personal communication 2009a), Keet (personal communication 2009b) and de Garine-Wichatitsky *et al.* (2010).

5 In 2006, a survey was also conducted in the LNP (Mozambique) and adjacent areas to detect BTB in buffalo and cattle populations frequently interacting inside and at the periphery of the park. No positive case was diagnosed. In 2008, another survey involving 489 cattle from four communal areas located at the periphery of GNP (Zimbabwe) did not find evidence of BTB infection in livestock (Gomo *et al.*, unpublished results). No positive case was diagnosed.

6 The frequency and perception of wildlife-borne diseases in livestock was investigated using three different sources of information during a study carried out in 2006–2007 in the Southeast Lowveld of Zimbabwe: (1) questionnaire survey on farmers' perception of the most frequent diseases affecting their livestock and the possible role played by wildlife in transmitting or maintaining these diseases (10 dip tanks attached to Malipati Animal Health Centre, Chiredzi District, $n = 431$ cattle keepers); (2) records of disease cases affecting livestock in the area of interest during the period 2006–2007, as reported to the District Veterinary Office in Chiredzi; (3) preliminary results of serological tests carried out on samples collected from cattle in the Malipati area during surveillance campaigns in 2007 and 2008 (PARSEL project, unpublished results).

7 Following the breakdown of dipping with acaricides during the liberation war, an estimated 1 million cattle died of tick-borne diseases during the late 1970s in Zimbabwe (Lawrence *et al.*, 1980). During the following years, most cattle populations in the communal lands of the country experienced a situation of 'enzootic stability' to some tick-borne diseases (e.g. babesiosis, Norval *et al.*, 1983), due to early immunization of young cattle challenged by infected ticks. This epidemiological situation should be promoted as a way to reduce the number of acaricide treatments and the overall impacts of tick-borne diseases (Norval, 1981), and it is likely that numerous cattle populations in the communal lands bordering protected areas of the country are indeed in a situation of endemic stability, as most dip-tank activity was disrupted during the period 2000–2009, with irregular or no supply of acaricide.

References

Anderson, E.C., Foggin, C., Atkinson, M., Sorensen, K.J., Madekurozva, R.L. and Nqindi, J. (1993) 'The role of wild animals, other than buffalo, in the current epidemiology of foot-and-mouth disease in Zimbabwe', *Epidemiology and Infection*, vol. 111, no. 3, pp. 559–563

Anderson, E.C. and Rowe, L.W. (1998) 'The prevalence of antibody to the viruses of bovine virus diarrhoea, bovine herpes virus 1, Rift Valley fever, ephemeral fever and bluetongue and to *Leptospira sp* in free-ranging wildlife in Zimbabwe', *Epidemiology and Infection*, vol. 121, no. 2, pp. 441–449

Anthony, B. (2007) 'The dual nature of parks: attitudes of neighbouring communities towards the Kruger National Park, South Africa', *Environmental Conservation*, vol. 34, pp. 236–245

Anthony, B.P. (2006) 'A view from the other side of the fence: Tsonga communities and the Kruger National Park, South Africa', Central European University, Budapest

Baudron, F., Corbeels, M., Anderson, J.A., Sibanda, M. and Giller, K.E. (2011) 'Delineating the drivers of waning wildlife habitat: the predominance of cotton farming on the fringe of protected areas in the Mid Zambezi Valley, Zimbabwe', *Biological Conservation*, vol. 144, no. 5, pp. 1481–1493

Bengis, R.G. (2005) 'Transfrontier conservation area initiatives in Sub-Saharan Africa: some animal health challenges', in S.A. Osofsky, S. Cleaveland, W.B. Karesh, M.D. Kock, P.J. Nyhus, L. Starrand and A. Yang (eds), *Conservation and Development Interventions at the Wildlife/Livestock Interface: Implications for Wildlife, Livestock and Human Health*, IUCN, Gland and Cambridge, UK

Bengis, R.G., Kock, R.A. and Fischer, J. (2002) 'Infectious animal disease: the wildlife livestock interface', *OIE Revue Scientifique et Technique*, vol. 21, no. 1, pp. 53–65

Bengis, R.G., Kriek, N.P., Keet, D.F., Raath, J.P., de Vos, V. and Huchzermeyer, H.F. (1996) 'An outbreak of bovine tuberculosis in a free-living African buffalo (*Syncerus caffer* –sparrman) population in the Kruger National Park: a preliminary report', *Onderstepoort Journal of Veterinary Research*, vol. 63, no. 1, pp. 15–18

Bourn, D. and Blench, R. (1999) *Can Livestock and Wildlife Co-exist: An Interdisciplinary Approach*, Overseas Development Institute, London

Caron, A., Pfukenyi, D., Murwira, A., Njagu, C. and de Garine-Wichatitsky, M. (2010) 'Seasonal and spatial patterns of foot-and-mouth disease outbreaks in cattle at the wildlife/livestock interface in Zimbabwe with particular attention to south-east lowveld (1997–2006)', *FMD International Symposium and Workshop, 12–14 April, Melbourne, Australia, 2010*, Melbourne

Chamaillé-Jammes, S., Fritz, H., Valeix, M., Murindagomo, F. and Clobert, J. (2008) 'Resource variability, aggregation and direct density dependence in an open context: the local regulation of an African elephant population', *Journal of Animal Ecology*, vol. 77, no. 1, pp. 135–144

Chaminuka, P., McCrindle, C.M.E and Udo, H.M.J. (in press) 'Cattle farming at the wildlife/livestock interface: assessment of costs and benefits adjacent to Kruger National Park, South Africa', *Society and Natural Resources.*

Chaparro, F., Lawrence, J.V., Bengis, R.G. and Myburgh, J.G. (1990) 'A serological survey for brucellosis in buffalo (*Syncerus caffer*) in the Kruger National Park', *Journal of the South African Veterinary Association*, vol. 61, no. 3, pp. 110–111

Condy, J.B. (1979) 'A history of foot-and-mouth disease in Rhodesia', *Rhodesia Veterinary Journal*, vol. 10, pp. 2–10

Cumming, D.H.M. (1981) 'The management of elephants and other large mammals in Zimbabwe', in P.A. Jewel, S. Holtand and D. Hart (eds), *Problems in Management of Locally Abundant Wild Mammals*, Academic Press, New York

Cumming, D.H.M. (2008) 'Large scale conservation planning and priorities for the Kavango-Zambezi Transfrontier Conservation Area', report for Conservation International, Washington

Cumming, G.S., Cumming, D.H.M. and Redman, C. (2006) 'Scale mismatches in social-ecological systems: causes, consequences and solutions', *Ecology and Society*, vol. 11, pp. 14–34

Davies, F.G. (1975) 'Observations on the epidemiology of Rift Valley fever in Kenya', *Journal of Hygiene*, vol. 75, no. 2, pp. 219–230

Davies, F.G. (1982) 'Observations on the epidemiology of lumpy skin disease in Kenya', *Journal of Hygiene*, vol. 88, no. 1, pp. 95–102

De Lisle, G.W., Bengis, R.G., Schmitt, S.M. and O'Brien, D.J. (2002) 'Tuberculosis in free-ranging wildlife: detection, diagnosis and management', *OIE Revue Scientifique et Technique*, vol. 21, no. 2, pp. 317–334

Dube, T., Murwira, A., Caron, A. and Garine-Wichatitsky, M. de (2010) 'Preliminary results on the permeability of veterinary fences to buffalo (*Syncerus caffer)* and cattle in Gonarezhou National Park, Zimbabwe', in K. Ferguson and J. Hanks (eds), *A Review of the Environmental, Social and Economic Impacts of Game and Veterinary Fencing in Africa with Particular Reference to the Great Limpopo and Kavango-Zambezi Transfrontier Conservation Areas*, Mammal Research Institute, Pretoria

Dunham, K.M. (2002) 'Aerial census of elephants and other large mammals in the northwest Matebeleland, Zimbabwe: 2001', *WWF-SARPO Occasional Paper Number 6*, WWWF-SARPO, Harare

Dunham, K.M., Mackie, C.S., Musemburi, O.C., Zhuwau, C., Mtare, T.G., Taylor, R.D. and Chimuti, T. (2007) 'An aerial census of elephants and other large mammals in northwest Matebeleland, Zimbabwe', *WWF-SARPO Occasional Paper Number 19*, WWWF-SARPO, Harare

Evans, A., Gakuya, F., Paweska, J.T., Rostal, M., Akoolo, L., Van Vuren, P.J., Manyibe, T., Macharia, J.M., Ksiazek, T.G., Feikin, D.R., Breiman, R.F. and Kariuki Njenga, M. (2008) 'Prevalence of antibodies against Rift Valley fever virus in Kenyan wildlife', *Epidemiology and Infection*, vol. 136, no. 9, pp. 1261–1269

Ferguson, K. and Hanks, J. (2010) *A Review of the Environmental, Social and Economic Impacts of Game and Veterinary Fencing in Africa with Particular Reference to the Great Limpopo and Kavango-Zambezi Transfrontier Conservation Areas*, Mammal Research Institute, Pretoria

Garine, I. de (1993) 'Culture, seasons and stress in two traditional African cultures (Massa and Mussey)', in S.J. Ulijaszek and S.S. Strickl (eds), *Seasonality and Human Ecology*, Cambridge University Press, Cambridge

Garine, I. de (2005) 'Nourriture de brousse chez les Muzey et les Masa du Nord-Cameroun', in C. Raimond, E. Garine and O. Langlois (eds), *Ressources vivrières et choix alimentaires dans le bassin du lac Tchad*, IRD Editions, Paris

Garine, I. de and Garine-Wichatitsky, M. de (1999) 'Providing impala meat *(Aepyceros melampus)* to local populations in Nyaminyami (Omay, Zimbabwe): nutritional and sociocultural aspects of man/wildlife interactions', *Revue Bois et Forets des Tropiques*, vol. 262, no. 4, pp. 61–76

Garine-Wichatitsky, M. de, Caron, A., Gomo, C., Foggin, C., Dutlow, K., Pfukenyi, D., Lane, E., Le Bel, S., Hofmeyr, M., Hlokwe, T. and Michel, A. (2010) 'Spread of wildlife bovine tuberculosis across international boundaries of the Great Limpopo Trans Frontier Conservation Area', *Emerging Infectious Diseases*, vol. 16, no. 5, pp. 884–885

Godfroid, J. (2002) 'Brucellosis in wildlife', *OIE Revue Scientifique et Technique*, vol. 21, no. 2, pp. 277–290

Gomo, C. (2010) 'A survey of Tuberculosis and Brucellosis in wild and domestic animals in the South-East Lowveld of Zimbabwe', MPhil thesis, University of Zimbabwe, Harare

Grahn, R. and Leyland, T. (2005), 'Complementarity between community-based animal health delivery systems and community-based wildlife management? An analysis of experiences linking animal health to conflict management in pastoralist areas of the horn of Africa', in S.A. Osofsky, S. Cleaveland, W.B. Karesh, M.D. Kock, P.J. Nyhus, L. Starr and A. Yang (eds), *Conservation and Development Interventions at the Wildlife/Livestock Interface: Implications for Wildlife, Livestock and Human Health*, IUCN, Gland, pp. 133–139

Grobler, D.G., Michel, A.L., De Klerk, L.M. and Bengis, R.G. (2002) 'The gamma-interferon test: its usefulness in a bovine tuberculosis survey in African buffaloes (*Syncerus caffer*) in the Kruger National Park', *Onderstepoort J Vet Res*, vol. 69, no. 3, pp. 221–227

Grootenhuis, J.G. (1988) 'The role of wildlife in the epidemiology of cattle theileriosis', in T. Dolan (ed.), *Theileriosis in Eastern, Central and Southern Africa*, ILRAD, Nairobi

Hoare, R. (1999) 'Determinants of human–elephant conflict in a land-use mosaic', *Journal of Applied Ecology*, vol. 36, pp. 689–700

Huchzermeyer, H.F.K.A., Brückner, G.K., Van Heerden, A., Kleeberg, H.H., van Rensburg, I.B.J., Koen, P. and Loveday, R.K. (1994) 'Tuberculosis', in J.A.W. Coetzer, G.R. Thomson and R.C. Tustin (eds), *Infectious Diseases of Livestock with Special Reference to Southern Africa*, Oxford University Press, Oxford

Hugh-Jones, M.E. and Vos, V. de (2002) 'Anthrax and wildlife', *OIE Revue Scientifique et Technique*, vol. 21, no. 2, pp. 359–383

Jori, F., Vosloo, W., Du Plessis, B., Bengis, R.G., Brahmbhatt, D., Gummow, B. and Thomson, G.R. (2009) 'A qualitative risk assessment of factors contributing to foot and mouth disease cattle outbreaks along the western boundary of Kruger National Park', *OIE Revue Scientifique et Technique*, vol. 29, no. 3, pp. 917–931

Knobel, D.L., Cleaveland, S., Coleman, P.G., Fèvre, E.M., Meltzer, M.I., Miranda, M.E.G., Shaw, A., Zinsstag, J. and Meslin, F.X. (2005) 'Re-evaluating the burden of rabies in Africa and Asia', *Bulletin of the World Health Organization*, vol. 83, no. 5, pp. 360–368

Lamarque, F., Anderson, J., Fergusson, R., Lagrange, M., Osei-Owusu, Y. and Bakker, L. (2009) *Human–Wildlife Conflict. Causes, Consequences and Management Strategies*, FAO, Rome

Lawrence, J.A., Foggin, C.M. and Norval, R.A. (1980) 'The effects of war on the control of diseases of livestock in Rhodesia (Zimbabwe)', *Veterinary Record*, vol. 107, no. 4, pp. 82–85

Madsen, M. and Anderson, E.C. (1995) 'Serologic survey of Zimbabwean wildlife for brucellosis', *Journal of Zoo and Wildlife Medicine*, vol. 26, no. 2, pp. 240–245

Marange, R. and Marimwe, M.C. (2008) 'A retrospective study on transmissible diseases between wildlife and domestic livestock in the wildlife/livestock interface areas of Hwange district', BVSc thesis, University of Zimbabwe, Harare

Marks, S.A. (1976) *Large Mammals and a Brave People. Subsistence Hunters in Zambia*, University of Washington Press, Seattle

Metcalfe, S. and Kepe, T. (2008) '"Your elephant in our land." The struggle to manage wildlife mobility on Zambian communal land in the Kavango-Zambezi Transfrontier Conservation Area', *Journal of Environment and Development*, vol. 17, no. 2, pp. 99–117

Michel, A.L., Bengis, R.G., Keet, D.F., Hofmeyr, M., Klerk, L.M. de, Cross, P.C., Jolles, A.E., Cooper, D., Whyte, I.J., Buss, P. and Godfroid, J. (2006) 'Wildlife tuberculosis in South African conservation areas: implications and challenges', *Veterinary Microbiology*, vol. 112, pp. 91–100

Michel, A.L., Müller, B. and Helden, P. van (2009) '*Mycobacterium bovis* at the animal–human interface: a problem, or not?' *Veterinary Microbiology*, vol. 133, no. 4, pp. 335–343

Muir-Leresche, K. and Nelson, R. (2001) 'Managing Africa's wildlife', *PERC Reports*, vol. 19, no. 3, pp. 7–17

Norval, R.A.I. (1981) 'A re-assessment of the role of dipping in the control of tick-borne disease in Zimbabwe', *Proceedings of the International Tick Biology and Control*, Grahamstown

Norval, R.A.I., Fivaz, B.H., Lawrence, J.A. and Brown, A.F. (1984) 'Epidemiology of tick-borne diseases of cattle in Zimbabwe. II. Anaplasmosis', *Tropical Animal Health and Production*, vol. 16, no. 2, pp. 63–70

Norval, R.A.I., Fivaz, B.H., Lawrence, J.A. and Daillecourt, T. (1983) 'Epidemiology of tick-borne diseases of cattle in Zimbabwe. I. Babesiosis.' *Tropical Animal Health and Production*, vol. 15, no. 2, pp. 87–94

O'Connell-Rodwell, C.E., Rodwell, T., Rice, M. and Hart, L.A. (2000) 'Living with the modern conservation paradigm: can agricultural communities coexist with elephants? A five-year case study in East Caprivi, Namibia', *Biological Conservation*, vol. 93, no. 3, pp. 381–391

OIE (2003) *The Terrestrial Animal Health Code*, Paris

Osborn, F.V. (2004) 'Seasonal variation of feeding patterns and food selection by crop-raiding elephants in Zimbabwe', *African Journal of Ecology*, vol. 42, pp. 322–327

Osofsky, S.A., Cleaveland, S., Karesh, W.B., Kock, M.D., Nyhus, P.J., Starr, L. and Yang, A. (2005) *Conservation and Development Interventions at the Wildlife/Livestock Interface: Implications for Wildlife, Livestock and Human Health*, UICN, Gland and Cambridge, UK

Parker, G.E., Osborn, F.V., Hoare, R.E. and Niskanen, L.S. (2007) *Human Elephant Conflict Mitigation: A Training Course for Community-Based Approaches in Africa. Participant's Manual*, IUCN Species Survival Commission, African Elephant Specialist Group, Livingstone

Parker, G.E. and Osborn, F.V. (2001) 'Dual season crop damage by elephants in the Eastern Zambezi Valley, Zimbabwe', *Pachyderm*, vol. 30, pp. 49–56

Paumgarten, F. (2006) 'The role of non-timber forest products as safety-nets: a review of evidence with a focus on South Africa', *GeoJournal*, vol. 64, no. 3, pp. 189–197

Pauw, K. (2005) 'Profile of the Limpopo Province: demographics, poverty, inequality, and unemployment', *PROVIDE Project, Background Paper Series*, no. 15607. Available online: http://ageconsearch.umn.edu/bitstream/15607/1/bp050009.pdf (accessed 24 August 2011)

Paweska, J.T., Vuren, P.J. van, Kemp, A., Buss, P., Bengis, R.G., Gakuya, F., Breiman, R.F., Njenga, M.K. and Swanepoel, R. (2008) 'Recombinant nucleocapsid-based ELISA for detection of IgG antibody to Rift Valley fever virus in African buffalo', *Veterinary Microbiology*, vol. 127, no. 1–2, pp. 21–28

Du Plessis, B. (2004) 'Annual Report 2003', Animal Health Services, Ehlanzeni District, Nelspruit, Mpumalanga Province, South Africa

Du Plessis, B. (2005) 'Annual Report 2004', Animal Health Services, Ehlanzeni District, Nelspruit, Mpumalanga Province, South Africa

Renwick, A.R., White, P.C. and Bengis, R.G. (2007) 'Bovine tuberculosis in Southern African wildlife: a multi-species host-pathogen system', *Epidemiology and Infection*, vol. 135, no. 4, pp. 529–540

Rihoy, E., Chirozva, C. and Anstey, S. (2007) '"People are not happy." Speaking up for adaptive natural resource governance in Mahenya. Evolution and resilience in the face of adversity, or another case of community based natural resource management (CBNRM) in crisis?' *Land Reform and Agrarian Change in Southern Africa Occasional Paper Series*, University of the Western Cape, Cape Town

Sabeta, C.T., Bingham, J. and Nel, L.H. (2003) 'Molecular epidemiology of canid rabies in Zimbabwe and South Africa', *Virus Research*, vol. 91, pp. 203–211

Scholes, R., Carruthers, J., Ararde, R. van, Kerley, G., Twine, W., Grobler, D., Bertschinger, H., Grant, R., Slotow, R., Lotter, H., Blignaut, J., Hopkinson, L. and Biggs, H. (2007) 'Summary for policy makers: assessment of South African elephant management', in R. Scholes and K.G. Mennell (eds), *Elephant Management: A Scientific Assessment of South Africa*, Witswatersrand University, Johannesburg

Seekings, J. (2000) 'Visions of society: peasants, workers and the unemployed in a changing South Africa', *Journal for Studies in Economics and Econometrics*, vol. 24, pp. 53–71

Sitati, N.W., Walpole, M.J. and Leader-Williams, N. (2005) 'Factors affecting susceptibility of farms to crop raiding by African elephants: using predictive model to mitigate conflict', *Journal of Applied Ecology*, vol. 42, pp. 1175–182

Sitati, N.W., Walpole, M.J., Smith, R.J. and Leader-Williams, N. (2003) 'Predicting spatial aspects of human-elephant conflicts', *Journal of Applied Ecology*, vol. 40, pp. 667–677

Spenceley, A. (2005) 'Nature-based tourism and environmental sustainability in South Africa', *Journal of Sustainable Tourism*, vol. 13, pp. 136–170

Taylor, R. (2009) 'Community based natural resource management in Zimbabwe: the experience of CAMPFIRE', *Biodiversity and Conservation*, vol. 18, no. 10, pp. 2563–2583

Taylor, R.D. and Martin, R.B. (1987) 'Effects of veterinary fences on wildlife conservation in Zimbabwe', *Environmental Management*, vol. 11, no. 3, pp. 327–334

Taylor, R.D. and Murphree, M.W. (2007) 'Case studies on successful Southern African NRM inititiatives and their impacts on poverty and governance. Zimbabwe: Masoka and Gairez', IRG, Washington DC

Thomson, G.R. (2008) 'A short overview of regional positions on foot-and-mouth disease control in Southern Africa', in *Transboundary Animal Disease and Market Access: Future Options for the Beef Industry in Southern Africa*, Institute of Development Studies, Brighton

Thomson, G.R. and Bastos, A.D.R. (2004), 'Foot-and-mouth disease', in J.A.W. Coetzer and R.C. Tustin (eds), *Infectious Diseases of Livestock*, Oxford University Press, Oxford

Thomson, G.R., Tambi, E.N., Hargreaves, S.K., Leyland, T.J., Catley, A.P., Klooster, G. van and Penrith, M.L. (2004) 'International trade in livestock and livestock products: the need for a commodity-based approach', *Veterinary Record*, vol. 155, pp. 429–433

Thomson, G.R., Vosloo, W. and Bastos, A.D.R. (2003) 'Foot-and-mouth disease in wildlife', *Virus Research*, vol. 91, no. 1, pp. 145–161

Twine, W. and Magome, H. (2008), 'Interactions between elephants and people', in R. Scholes and K. Mennell (eds), *Elephant Management: A Scientific Assessment for South Africa*, Wittswatersrand University Press, Johannesburg

Vincent, V. and Thomas, R.G. (1960) *An Agricultural Survey of Southern Rhodesia. Part 1: Agro Ecological Survey*, Government Printer, Salisbury, Rhodesia

Vosloo, W., Bastos, A.D., Sangare, O., Hargreaves, S.K. and Thomson, G.R. (2002a) 'Review of the status and control of foot and mouth disease in sub-Saharan Africa', *OIE Revue Scientifique et Technique*, vol. 21, no. 3, pp. 437–449

Vosloo, W., Bastos, A.D., Sahle, M., Sangare, O. and Dwarka, R.M. (2005) 'Virus topotypes and the role of wildlife in foot and mouth disease in Africa.' in S.A. Osofsky, S. Cleaveland, W.B. Karesh, M.D. Kock, P.J. Nyhus, L. Starr and A. Yang (eds), *Conservation and Development Interventions at the Wildlife/Livestock Interface: Implications for Wildlife, Livestock and Human Health*, IUCN, Gland

Vosloo, W., Boshoff, K., Dwarka, R. and Bastos, A.D. (2002b) 'The possible role that buffalo played in the outbreaks of foot-and-mouth disease in South Africa', *Annals of the New York Academy of Science*, vol. 969, pp. 187–190

Zihumo, T. and Mapuvire, T. (2006) 'A retrospective study on the dynamics of foot-and-mouth disease in cattle in Zimbabwe for the period January 2000 to December 2006', BVSc thesis, University of Zimbabwe, Harare

9 Land- and natural resource-based livelihood opportunities in TFCAs

David H.M. Cumming, Vupenyu Dzingirai and Michel de Garine-Wichatitsky

Introduction

Transboundary parks in Southern Africa[1] are a relatively recent phenomenon. The Kalahari Gemsbok National Park in South Africa and the Kgalagadi National Park in Botswana shared a common boundary for 40 years before being formally merged into the Kgalagadi Transfrontier Park in May 2000. Mana Pools National Park in Zimbabwe and the Southern Zambezi National Park in Zambia have shared a boundary for 27 years, and Kruger and Limpopo National Parks have shared an international boundary for about 10 years. The recent move to link a range of different types of protected areas, that may be near each other but do not necessarily share common boundaries, is indeed a very new initiative. The concept began to receive serious attention in Southern Africa in the 1990s. From a conservation perspective, transfrontier conservation areas (TFCAs) promise to link protected areas of various types across large landscapes, both within and between countries, by providing an intervening matrix in which animals can move between otherwise ecologically isolated protected 'islands'. From the perspective of people living on the boundaries of national parks in the matrix between protected areas, i.e. on the 'the edge', and within a TFCA, a wide range of impacts on their livelihoods emerge. These, often negative, impacts have been explored in the earlier chapters of this book. This chapter examines the opportunities available to people living on the edge and within TFCAs in Southern Africa.

By opportunities, we mean the land and natural resource use options that may exist within TFCAs that might now, or in the future, generate wealth and improved livelihoods for those living within these areas. These opportunities vary considerably, depending on the country, the resource base and the legal and policy frameworks governing access to, and use of, natural resources. They also depend on human population densities, and the social and economic resources (social and physical capital) in the respective TFCAs.

There are now some 13 TFCAs under active development in Southern Africa (Figure 9.1). They are expected to meet conservation, development and political goals (see Chapter 2). With rapidly growing human populations increasingly occupying marginal areas the pressures on natural resources are mounting by the day. There is an urgent need to establish the policy and incentive frameworks across the region that will allow TFCAs to meet both the needs of people and the conservation of natural resources and biodiversity.

The boundaries of TFCAs, and the areas they encompass, are, in contrast to national parks, still mostly ill-defined. They vary in size from the Kavango–Zambezi (or KAZA TFCA) of *c.*400,000 km^2 that straddles five countries to the Kasungu–Lukusuzi TFCA in Zambia-Malawi of *c.*2,316 km^2.

This chapter deals with the central question of why, for the most part, people living within TFCAs are not reaping the potential benefits of wildlife and other natural resources in their areas. We first outline the land tenure and environmental characteristics of the TFCAs being developed in Southern Africa to provide a broad context in which enterprises based on indigenous fauna and flora may be developed to sustain livelihoods for those living on the edge. The major tourist attractions and enterprise opportunities in relation to the 13 Southern African TFCAs are briefly summarised and discussed before providing several specific examples of existing natural resource based developments within TFCAs in South Africa, Zimbabwe and Namibia. Having provided the broad context for natural resource-based enterprise development and some specific current examples, we then consider the role of legal and policy constraints that inhibit improved benefit streams to those living on the edge, together with some specific examples of likely potential returns from CBNRM. The potential influence of scientific, technical and cultural exchanges across international boundaries is briefly examined. In conclusion, we consider the necessary conditions for moving from opportunities to implementation and outline four strategies that may contribute to improved natural resource management and the flow of benefits to those living on the edge, as well as to conservation of the region's biodiversity.

Land tenure and environmental features of Southern African TFCAs

Size, land tenure and fragmentation

Formal treaties between countries in Southern Africa have been signed for the establishment of three transfrontier national parks (Kgalagadi, Ai-Ais/Richtersveld and Great Limpopo). Memoranda of Understanding (MOUs) have been signed for the development of seven TFCAs, and the MOUs for a further three TFCAs are under development. In terms of the areas of land involved, three very large TFCAs, each covering more than 80,000 km^2, are being developed. A further three cover between 20,000 and 40,000 km^2, and the remaining eight TFCAs vary between 2,000 and 13,000 km^2 (Table 9.1). One of the TFCAs, the Kavango-Zambezi (KAZA), involves five countries; the Great Limpopo TFCA (GLTFCA) and the Greater Mapungubwe TFCA each involve three countries, while the remaining nine TFCAs involve only two countries. The degree of coherence in terms of the number of units, and, consequently, the extent of edges between

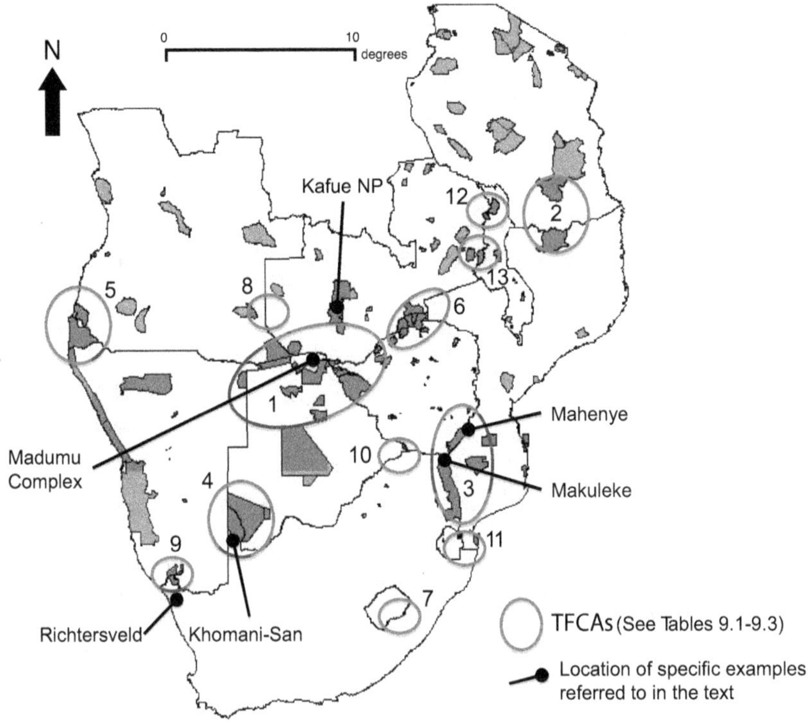

Figure 9.1 Map of Southern Africa showing major protected areas (shaded) and TFCAs that are being developed. The numbers and names and additional details of each TFCA are provided in Table 9.1. Particular project areas covered in the text of this chapter are shown by dots and a short name for the area (e.g. Mahenye).

protected and non-protected areas, varies considerably between TFCAs. Three areas form unified, largely unfragmented protected areas, namely, Kgalagadi TFP, Iona-Skeleton Coast TFCA, and Liuwa Plain TFCA. The remaining areas comprise collections of protected areas embedded within intervening unprotected land – mostly communal land. The length of 'edge' in these areas is, as a result of fragmentation, greatly increased (Figure 9.1 and Table 9.1) and the number of land category units within the TFCAs (Table 9.1) provides an indication of the extent of land use boundaries and resulting opportunities, risks and conflicts that can potentially arise in these areas. In some cases the land involved is entirely state land, e.g. Kgalagadi and Iona (Figure 9.1) but in others such the KAZA TFCA only 22 per cent is contained in state-protected areas (Table 9.1). In nearly all TFCAs small-scale crop and livestock farming form a major component of the matrix (see Chapter 2).

Table 9.1 The size, land tenure regimes and land uses of thirteen TFCAs in Southern Africa. A summary of (a) the number of countries involved in each TFCA and its area, (b) the major land tenure regimes within each TFCA and the approximate number of designated units of each type falling within the TFCA, which provides an indication of complexity, (c) the approximate percentage of the overall area of land within the TFCA that falls under different land tenure regimes, and (d) an indication of the major land uses outside state protected areas (STAs) but within the TFCA (+ = present; – = not present or minimally present; ? = uncertain or not known).

TFCA			*Major land tenure regimes (no. of units)*								*% Land*					*Land uses outside SPAs*				
	No. countries	*Area (km²)*	*National parks*	*Safari areas and hunting blocks*	*Forest reserves*	*Communal land wildlife areas*	*Private land conservancies*	*Community conservancies*	*Private farm land*	*Communal land*	*State-protected areas*	*Communal land wildlife areas*	*Private land conservancies*	*Private farm land*	*Communal land*	*Crops*	*Stock*	*Irrigation*	*Wildlife (non-consumptive)*	*Wildlife (consumptive)*
1 Kavango-Zambezi	5	400,000	17	4	14	14	1	20+	✓	✓	22	25	1	2	52	✓	✓	?	✓	✓
2. Niassa-Selous	2	96,200	2	??	–	–	–	–	–	✓	91	–	–	–	9	✓	✓	?	✓	✓
3. Great Limpopo	3	87,000	5	2	–	3	20+	3	✓	✓	53	34	✓	5	8	✓	✓	✓	✓	✓
4. Kgalagadi TFP	2	37,256	3	–	–	–	–	1	2	–	100	–	–	–	–	–	–	–	✓	–
5. Iona-Skeleton Coast	2	32,000	2	–	–	–	–	–	–	–	100	–	–	–	–	–	–	–	✓	–
6. Mana Pools-Lower Zambezi	2	25,000	4	3	–	2	–	–	–	✓	80	20	–	–	–	✓	✓	–	✓	✓
7. Drakensberg-Maloti	2	13,000	4	–	1	–	?	?	?	✓	100?	?	?	?	?	–	✓	–	✓	–
8. Liuwa Plain	2	10,000?	1	1	–	1	–	–	–	?	35	65	–	–	–	✓	✓	–	✓	✓
9. Ai-Ais/Richtersveld	2	6.681	2	–	–	–	–	1	–	✓	76	–	–	24	–	–	✓	–	✓	–
10. Greater Mapungubwe	3	4,872	2	1	–	–	6?	–	✓	✓	50	–	35	5	10		✓	✓	✓	
11. Lebombo	2	4,195	3	–	?	–	?	?	3+	✓	?	?	✓	✓	✓	✓	✓	✓	✓	✓
12. Nyika-Vwaza Marsh	2	4,134	3	–	4	1	–	1	–	✓	70	?	–	–	?	✓	✓	–	✓	–
13. Kasungu-Lukusuzi	2	2,316	2	–	–	–	–	–	–	✓	60	90?	–	–	10?	✓	✓	–	✓	✓

Major land tenure categories:
State protected areas (SPAs) include: national parks, safari areas and forest reserves.
Designated communal land wildlife areas include: game management areas (Zambia), wildlife management areas (Botswana), CAMPFIRE areas (Zimbabwe) and coutadas (hunting blocks) in Angola and Mozambique.
Freehold or leasehold land may include: private land conservancies, game ranches and private/commercial farm land (not under wildlife).
Traditional communal farmland (communal land) may include: community conservancies (Namibia), wilderness areas (Zimbabwe) and communal farming land.

Climate and agricultural potential of TFCAs

Climatically and topographically, the TFCAs range from cool mountainous areas, e.g. the Drakensburg-Maloti of South Africa and Lesotho and the Nyika plateau of Malawi and Zambia, to hot, low-lying and coastal areas such as the Iona, GLTFCA and the Lebombo (Table 9.2). They also range from regions receiving more than 1,200 mm of rain per annum (e.g. Niassa-Selous) to desert areas on the west coast with an annual precipitation of less than 100 mm. Agricultural potential also varies from areas with rich soils and reliable rainfall (e.g. Lebombo) to areas of nutrient-poor soils but adequate rain (most of KAZA and Niassa-Selous), or areas of nutrient-rich soils but poor and erratic rainfall (e.g. GLTFCA). A range of animal and human diseases are also closely linked to climate and habitats (Table 9.2). Arid areas experience minimal exposure to diseases (e.g. Ai-Ais/Richtersveld) while those in low-lying areas such as the Zambezi Valley carry malaria and trypanosomiasis of livestock (and occasionally humans). These features, together with human population density and levels of infrastructure development (Table 9.2), are important because they play a central role in determining the land use and livelihood options, tourism opportunities and trade-offs for people living within TFCAs on the edge of protected areas. A potentially important but as yet little explored opportunity involves the provision of ecosystem services from protected areas and TFCAs in general, e.g. water provision, wetlands that mitigate floods and intact woodlands for carbon sequestration, and the provision of a range of non-timber forest products that support rural households. The potential to provide some of these opportunities is indicated in Table 9.3.

Natural resource endowment, infrastructure and assets

The endowment of the 13 Southern African TFCAs in terms of their tourist attractions varies greatly, with only a few providing four or more of the seven major attractions listed in Table 9.3. Similar considerations apply to the provision of ecosystem services within and beyond their boundaries. Of the eight types of natural resource-based enterprises that could provide development opportunities for those living within TFCAs, some, such as hotels and lodges, camping trails and trophy hunting, are widely applicable (Table 9.3). Others, such as timber extraction, commercial benefits from non-timber forest products, and the development of multi-species systems that involve the management of livestock and large wild herbivores on the same range (as for example in the Ol Pejeta Conservancy[2] in Kenya) appear to be more limited in their application (Table 9.3). Non-consumptive tourism is a potential land use in all TFCAs but consumptive use of wildlife, e.g. sport hunting and cropping, currently occurs only in three of the largest TFCAs (Table 9.3). Harvesting of indigenous hardwoods is a commercial activity within the Kavango-Zambezi and Niassa-Selous TFCAs. Harvesting of non-timber forest products occurs in all TFCAs. In many cases there is potential for sustainable commercial harvesting and marketing of high-valued goods such as medicinal plants and oils (Bennett, 2006; Le Breton and Nemarundwe, 2009),

Table 9.2 Climatic indicators, soil status, temperature, population density, infrastructure development, and important diseases that occur within Southern African TFCAs

TFCA	Climate and soils				Population and development		Important diseases							
	Area km²	*Mean annual rainfall (mm)*	*Altitude (m)*	*Soils*	*Population density (km²)*	*Infrastructure*	*Malaria*	*FMD*	*BTB*	*Theileriosis*	*Trypanosomiasis**	*CBPP*	*ASF*	*Rabies*
1 Kavango-Zambezi	400,000	400–1,000	300–1,500	D/E	5	Low	+	+	+	+	+	+	+	+
2 Niassa-Selous	96,200	1,000–1,400	100–1,200	D	5–25	Very low	+	+	?	+	+	–	+	+
3 Great Limpopo	87,000	200–800	100–600	E/D	5–250	High	+	+	+	+	+	–	+	+
4 Kgalagadi TFP	37,256	200–600	600–900	D	5	Low	–	–	–	–	–	–	–	+
5 Iona-Skeleton Coast	32,000	< 200	0–400	E	< 1	Very low	–	–	–	–	–	–	–	+
6 Mana Pools-Lower Zambezi	25,000	600–800	300–1,000	D/E	5–25	Low	+	+	–	+	+	–	?	+
7 Drakensberg-Maloti	13,000	800–1,000	1,300–5,600	?	0–250	High/low	–	–	–	–	–	–	–	+
8 Liuwa Plain	10,000?	800–1,000	*c.* 1,000	D	5	Very low	+	+	?	+	+	?	+	+
9 Ai-Ais/Richtersveld	6,681	< 200	600–900	E	5	Low	–	–	+	–	–	–	–	+
10 Greater Mapungubwe	4,872	200–400	500–800	E	5–25	High/low	+	+	–	+	–	–	–	+
11 Lebombo	4,195	400–1,000	100–500	E/D	5–250	High	–	+	+	+	+	–	?	+
12 Nyika-Vwaza Marsh	4,134	1,200 > 1,400	600–2,600	D	5–250	Low	?	+	?	+	+	–	?	+
13 Kasunga-Lukusuzi	2,316	800–100	1,000–1,500	D	5–250	Low	+	+	?	+	+	–	?	+

Notes:
* Trypanosomiasis includes nagana, which affects domestic livestock, and human sickness, which is presently limited to occasional cases in the Zambezi Valley.
For soils, D = moist dystrophic (mostly sandy nutrient-poor soils) and E = Eutrophic nutrient-rich soils. For diseases: FMD = Foot and Mouth Disease, BTB = Bovine tuberculosis, CBPP = Contagious bovine pleuropneumonia and ASF = African swine fever. (+ = present, ? = status uncertain, – = probably absent)

Table 9.3 Main tourist attractions and natural resources-based development and livelihood opportunities in Southern African TFCAs

TFCA		*Main tourist attractions*						*Development and livelihood opportunities*											
		Wetlands	*Big five*	*Scenic*	*Species diversity*	*Wilderness*	*Developed*	*Accessible*	*Hotels & lodges*	*Camping & trails*	*Trophy hunting*	*Citizen hunting*	*Game cropping*	*Timber*	*NTFPs*	*Multi-species ranching*	*Water & wetlands*	*Carbon sequestration*	*Biodiversity*
1	Kavango-Zambezi	✓	✓	✓	✓			✓	✓	✓	✓	✓		✓	✓	✓	✓	✓	✓
2	Niassa-Selous	✓	✓		✓	✓			✓		✓	✓	✓	✓	✓		✓	✓	✓
3	Great Limpopo		✓		✓		✓	✓	✓	✓	✓		✓		✓	✓			✓
4	Kgalagadi TFP					✓			✓							✓			
5	Iona-Skeleton Coast	✓		✓		✓				✓									
6	Mana Pools-Lower Zambezi		✓						✓	✓	✓								
7	Drakensberg-Maloti	✓		✓				✓									✓		
8	Liuwa Plain	✓				✓				✓	✓	✓					✓		
9	Ai-Ais/Richtersveld								✓										
10	Greater Mapungubwe		✓					✓	✓	✓	✓		✓		✓	✓			
11	Lebombo		✓	✓	✓		✓	✓	✓	✓	✓				✓		✓	✓	✓
12	Nyika-Vwaza Marsh	✓		✓					✓					✓	✓		✓		✓
13	Kasungu-Lukusuzi													✓	✓			✓	

Note: Water and wetlands, carbon sequestration and biodiversity fall within the category of ecosystem services provision for which payments can potentially be received. Only those areas that have high levels of endemic species or key populations of endangered species, or both, have been ticked.

'mopane worms' (Stack *et al.*, 2003) and even charcoal for export from bush encroached areas where grazing needs to be re-established.

In addition to the natural resources endowment of TFCAs, infrastructure and the provision of associated services play an important role in the potential for these areas to provide development and livelihood opportunities for those living on the edge. Because most TFCAs are situated in the peripheral zones of their respective countries (see Chapters 6 and 3), the infrastructure and services are mostly poorly developed and greatly constrain development options (Table 9.2). There is also an inherent conflict between infrastructure development and the continuing provision of attractive wilderness experiences for tourists. For example, Kruger National Park (KNP) has a well-developed infrastructure both within the park and on its periphery and is able to attract a million or more tourists a year, but in the perception of many it does not provide a true wilderness experience. On the other hand, Gonarezhou National Park is poorly developed, attracts fewer than 10,000 visitors a year but provides a sense of wilderness that attracts a different category of tourist. The plan for the Gonarezhou National Park seeks to maintain its wilderness character.

Legislation and social capital

National legislation and policies for community-based natural resource management contain many common elements across the region, such as support for the sustainable use of natural resources and for communities to benefit directly from the utilisation of wildlife and associated tourism developments (Jones 2008, 2009; Cirelli and Morgera 2010). However, resource access rights vary between countries as does the extent to which communities can enter into joint ventures, private sector partnerships and commercial exploitation of their resources. For the most part, central and local governments exact considerable taxes on the returns that communities derive from the commercial use of their wild natural resources. This issue is discussed later in this chapter.

Social capital within communities living on the edge varies greatly across the region and has been influenced, among others, by the extent to which colonial and post-colonial practices have impacted social cohesion (e.g. forced displacements), by the nature and duration of civil wars and the movements of refugees, and the extent to which national boundaries have fragmented communities (see Chapters 3 and 4).

Natural resource-based enterprise opportunities

The major focus of TNP and TFCA development has been on tourist developments centred on Southern Africa's charismatic large mammal fauna, particularly the 'big five', and wild areas (Table 9.3). Related factors attracting foreign, high fee-paying tourists are wetlands, perceived wilderness qualities, biodiversity and accessibility. Hotels, lodges, camping facilities and guided trails have been, and can be, developed in all TFCAs and TNPs (Table 9.3). Large areas of most TFCAs lack the scenery, habitat, large mammal populations and infrastructure to attract and support a high throughput of non-consumptive (game viewing/photographic) tourism. However, many of these less attractive areas can and do support viable trophy hunting operations. In some countries hunting by citizens, as opposed to high fee-paying visitors, is permitted and where game populations can support it, animals are cropped to provide game meat for local and urban markets. The provision of ecosystem services and payments to local communities for these services (e.g. water supplies and flood mitigation, carbon sequestration and biodiversity) has hardly been explored.

Despite the many existing constraints to the development of viable and sustainable community enterprises for those living on the edge, there are some examples in the region that provide a basis on which to further explore livelihood opportunities for those living in TFCAs. Selected examples from the region are briefly outlined in this section before considering the opportunities that might be developed, given appropriate polices and incentive structures.

Examples of opportunities being developed

Community land and wildlife-based tourism in Zimbabwe

In many communal lands on the edge of protected areas, there is land that has remained unsettled and little used for grazing livestock because of a lack of water and its remote or rugged nature. This has the potential to provide wildlife-related benefits to communities.

One of the earliest examples of a community–private sector tourism initiative 'on the edge', and one that preceded the official start of Zimbabwe's CAMPFIRE, involved the Mahenye community on the north-eastern boundary of the Gonarezhou National Park (GNP). A number of households of the Mahenye community were displaced in the 1960s from the Lundi River area that is now within the GNP (see Chapter 3). Some households, however, remained settled on an island of ~5 km^2 in the lower Save River. The settlement was frequently subjected to crop raiding by elephants and a local safari operator, well-known and respected by the community, suggested that they move off the island and use it as a safari, or trophy hunting area. This they did in 1982/83 and the area has since been used for trophy hunting. By 1987 the proceeds from hunting were used to build the first primary school in the ward and a grinding mill (Peterson, 1991). Further developments followed in the early 1990s. Land was leased to Southern Sun Hotels to build two luxury lodges: Mahenye Lodge on an island in the Save River and Chilo Lodge on the eastern bank of the Save River adjacent to the GNP, together with the establishment of a further wildlife area within the Mahenye Ward (Murphree, 1995). Despite many problems in governance and the distribution of benefits (Rihoy *et al.*, 2004) the Mahenye community continues to plan new developments based on their wildlife resources (Stockil, personal communication, 2010).

On the opposite side of GNP, the Naivasha area of approximately 300 km^2 (part of Matibi No. 2 Communal Land) was set aside as a wildlife area and enclosed within the GNP when the park was fenced as a Foot and Mouth Disease (FMD) control measure (Chapter 3). The Chiredzi Rural District Council leases the area as a hunting concession. The Malipati Safari Area was initially designated as a game reserve within the Matibi Communal Land in 1968 and managed by what was then Ministry of Internal Affairs, with the express purpose of providing benefits to the surrounding community. The area was incorporated into the Parks and Wildlife Estate after Independence in 1980 and gazetted as a Safari Area. Revenue derived from the lease of hunting was returned to the Chiredzi Rural District Council until the mid-1980s and thereafter was retained by National Parks. Later still the area was leased to the Rural District Council (RDC), which in turn could lease out the hunting rights. Community rights to the benefits generated by the Malipati area have still to be reinstated.

The Kanyurira Ward of 354 km^2, in the Mid-Zambezi Valley of Zimbabwe, is wedged between the Chewore and Dande Safari Areas. The greater proportion of the ward is occupied by the Zambezi escarpment and is unsuitable for agriculture. Settlement is confined to rich alluvial areas along the Angwa River at Masoka. The area falls within the tsetse belt, which precludes the keeping of cattle. The Masoka community, at the onset of CAMPFIRE in 1991, opted to enclose their village within a fenced area of 20 km^2 and to use the remainder of their ward for wildlife. They regarded their wildlife as their cattle and proceeds from safari hunting have allowed them to weather periods of drought when crops failed, to build a primary and a secondary school, to purchase and maintain a tractor to till their fields, and in some years to pay a household dividend (Figure 9.2). The community has, over a period of twenty years, grown from 65 to >160 households and they have extended their enclosed settled area but taken care to maintain a viable wildlife area. As in the case of Mahenye, they have faced governance issues and problems with human–wildlife conflict but continue to derive substantial benefits from a wildlife-based economy that is linked to the neighbouring protected areas adjoining their ward.

Land claims and contractual parks in South Africa

Before 1994 the South African government developed contractual arrangements between private landowners and the state for the joint management of conservation areas on private land that abutted existing or newly created national parks. During the 1980s, moves by the National Parks Board to develop a contractual park with the Richtersveld community were initiated. This resulted in the Richtersveld National Park of 1,624 km^2 being established in 1991 following a court action by the community to enforce the signing of an agreement that better met their objectives (Grossman and Holden, 2009). The park adjoins the Ais-Ais National Park in Namibia and now forms part of the Ais-Ais/Richtersveld Transfrontier Park.

Following the new dispensation in 1994, the Makuleke community filed a land claim for the area between the Livuvu and Limpopo Rivers from which 3,000 members of the clan were evicted in 1969. Their claim was successful and that

Figure 9.2 Chief Kanyurira of Masoka receiving a household dividend at a village gathering.

(Photo: David Cumming)

part of KNP was deproclaimed as national park and reproclaimed as a Contractual National Park in 1998. On the opposite side of the country, the Khomani San and Mier Communities lodged successful land claims against the Kgalagadi National Park, with each being granted title to 250 km^2 within the park (Chennels, 2001). These areas were similarly reproclaimed as Contractual National Parks with title and land rights resting with the respective communities.

The results of these initiatives in terms of opportunities and benefits to communities within the respective TFCAs have been mixed. In the case of the Makuleke community, the land was vested in the Community Property Association and they were able to realise returns of R2 million from hunting safaris in the initial two years of the co-management of the contractual park. They later granted leases for the development of lodges, which have resulted in a continuing return of revenue. The National Parks Board makes an annual payment of R0.5 per ha to the Richtersveld Community Trust as part of the agreement in establishing the national park. The Khomani San have full rights of access for their traditional practices to the 250 km^2 of land they own within the Kgalagadi Transfrontier Park. However, the poor resource base and remoteness of the arid areas of the northern Cape and the Kalahari resulted in meagre tourist revenues for the Richtersveld, Khomani San and Mier communities. These communities also experienced major difficulties in building the institutions required to be effective partners in the co-management arrangements needed to effect development and provide meaningful returns to communities (Robins, 2001; Steenkamp and Grossman, 2001; Reid *et al.*, 2010; Grossman and Holden, 2009). More recently the fortunes of the Khomani San and Mier have taken a turn for the better with the completion and opening of the 24 bed !Xaus Lodge in 2007 under a contract with Transfrontier Park Destinations. Occupancy and returns have grown and over a three-year period the enterprise has generated more than R5.1 million. Community members form more than 85 per cent of lodge staff, and the communities own the R11 million lodge.

Community conservancies in Namibia

Namibia has experienced a dramatic increase in the number of community wildlife conservancies, from a single pilot project in the 1980s to 59 registered conservancies covering 132,697 km^2 in 2009 (NACSO, 2010). The annual contributions from CBNRM to net national income grew from less than N\$10 million in 1997 to more than N\$266 million in 2007, a figure that included the value of wildlife stocks (NACSO, 2010). The value of benefits earned for communities in 2009 was N\$42,481,015, of which cash income amounted to N\$25.9 million, non-cash income amounted to N\$9.1 million and other income from CBNRM but not related to conservancies amounted to N\$7.5 million. The main sources of income for conservancies in 2009 were joint venture tourism, game meat distribution and trophy hunting (Table 9.4).

There is considerable variation between conservancies in the amounts they earn and the source of their earnings. Some conservancies derive their income almost entirely from joint venture tourism while the income of others is derived mainly from trophy hunting. The value of game meat increased more than 30-fold between 1999 and 2009 to N\$6.3 million (NACSO, 2010).

The Namibian conservancies are mostly situated in the arid north-west of the country in the Kunene District and in the north-east, including the Caprivi, where there are 15 smaller conservancies. The Caprivi conservancies are situated in the heart of the Kavango-Zambezi (KAZA) TFCA within an area that carries the world's largest elephant population. The Caprivi has three national parks and a forest area covering an area of 7,044 km^2 and 15 community wildlife conservancies

Table 9.4 Main sources of income to conservancies in Namibia in 2009 (NACSO, 2010)

Source of income	**N$ (millions)*	*%*
Joint venture tourism	20.0	57.1
Game meat distribution and own use of game	6.3	18.0
Trophy hunting	5.7	16.3
Non-timber forest products and crafts	1.8	5.2
Other (campsites, game sales, premium hunting)	1.2	3.4
Total	35.0	100

* During 2009 approximately N$7.0 was equal to 1US$.

varying in size from 150 to 400 km^2. Communities derive their revenue mostly from tourist lodges and safari hunting, e.g. Mayuni Conservancy, covering 169 km^2, earned approximately N$900,000 in 2007 of which 50 per cent was derived from joint venture tourism and 30 per cent from trophy hunting.

The Nyae-Nyae Conservancy in the north-east of Namibia provides a second example. The conservancy is partly bounded by the Khaudom National Park (3,800 km^2) to the north and the Botswana border to the east and the veterinary cordon fence to the south. The conservancy (formerly Bushmanland) covers an area of 9,030 km^2 and was registered in 1998. In 2006 overall income from wildlife and tourism was N$114,860. Of this almost half was expended on community benefits with the following breakdown: 20% on protecting water supplies (from elephant), 19 per cent on natural resource monitoring, 4.5 per cent on governance meetings and 55 per cent on cash dividends to 657 members. Each member received N$300 (Jones and Weaver, 2009). Wildlife populations in the conservancy increased from 2,000 head in 1998 to more than 6,000 in 2004, with major increases in the numbers of wildebeest, springbok, kudu, oryx and zebra (Jones and Weaver, 2009; Weaver and Skyer, 2005).

Developments within TFCAs elsewhere in the region have not been as promising. In Botswana, although high financial returns were received initially by community-based organisations with wildlife concessions, central government has increasingly exerted control over their operations (Nelson and Agrawal, 2008; Denkler, 2009). The result is declining interest in managing wildlife and mistrust of government agencies (Denkler, 2009; Rozemeijer, 2009). In Mozambique, CBNRM does not yet appear to have been developed within TFCAs. Attempts to develop CBNRM within the country have faltered through uncertainties in the interpretation and implementation of complex land laws and how they relate to access rights to natural resources (Norfolk and Tanner, 2007; Anstey 2008; Nhantumbo and Anstey, 2009).

Most national parks in Zambia are surrounded by game management areas (GMAs) that were intended to provide a buffer zone between national parks and agricultural areas. They are communal lands in which agriculture is practised and few have retained substantial wildlife populations. The Kafue National Park of 22,400 km^2 is the largest in Zambia and forms an important part of the KAZA TFCA. It is surrounded by nine GMAs that cover about 46,000 km^2. The administration of hunting is controlled by the Zambia Wildlife Authority (ZAWA), which retains 50 per cent of the revenue derived from trophy and lease fees. The balance is passed back through the ADMADE (Administrative Management Design) programme of which 47 per cent is earmarked for wildlife management (e.g. game scout salaries and patrols) and 53 per cent for community development. In 1994, for example, trophy hunting revenues in GMAs throughout Zambia generated US$1.3 million of which the central treasury received US$424,656, the then National Parks and Wildlife Service US$218,148, and ADMADE US$654,442, with US$304,915 being allocated to community development (Lewis and Alpert, 1997). At the time GMAs covered more than 140,000 km^2 and the returns for community development represented a return of US$2.2 per km^2 or 2 cents per ha. By 2009 this had risen to US$10.7 per km^2 or 11 cents per ha. With human population densities greater than 5 people per km^2 this level of return amounts to about US$10 per household per annum. Other than for a few premium GMAs (Bandyopadhyay and Tembo, 2009; Fernandez, 2009) there is little incentive for those living in the GMAs to conserve wildlife.

In summary, apart from a few exceptions, such as the community conservancies in Namibia, most of the community wildlife-based programmes aimed at improving livelihoods provide limited benefits to communities and only weak incentives to conserve wildlife. The major constraints to the development of effective natural resources-based enterprises centre on governance issues and policies.

Legal and policy constraints to benefits from natural resource management in Southern Africa

The legal framework for much of Southern Africa is Roman-Dutch law, within which wildlife is considered to be *res nullius*, i.e., it belongs to no one until captured or killed and is essentially a fugitive resource (see Chapter 8). The result is that states have claimed ownership of wildlife, forests and fisheries and exercised control over access rights to wild large mammals and timber in particular, through laws and regulations. When human populations were sparse and law enforcement relatively strong, the state was able to maintain some semblance of control over the use of wildlife and to protect the resource. However, that control has been steadily eroded over the last five decades. In recent years, and in many areas, natural resource use has tended to become 'open access'. The breakdown in control results from a scale mismatch between institutions and the resource (Cumming *et al.*, 2006). Innovative wildlife legislation in the 1960s and 1970s, in South Africa, Namibia and particularly in Zimbabwe (Child, 1995; Bond and Cumming, 2006), devolved the rights and responsibilities (legally 'appropriate authority') for the use of wildlife to landholders. The result was a rapid expansion of wildlife populations and the wildlife industry on land under freehold title. In the communal lands of Zimbabwe the then Ministry of Internal Affairs assumed appropriate authority for wildlife. With the advent of the CAMPFIRE, appropriate authority for wildlife in the communal lands was granted to RDCs participating in the programme. Following independence in Namibia, and the formation of community conservancies, full benefits from wildlife-based tourism were devolved to communities but government continued to retain control over quotas. The failure to devolve resource access rights, responsibilities and full benefits to an appropriate level in most communal lands within Southern Africa has resulted in a disconnect between the costs and benefits rural communities experience in living with wildlife. For the most part, they have little if any direct involvement in the management of wildlife, which is not seen to belong to them. The result is a scale mismatch between institutions and resource management (Cumming *et al.*, 2006) with people living in communal lands perceiving wildlife as a cost to be minimised rather a benefit to be nurtured. As Mr Mudemi, one-time CAMPFIRE Officer for Beitbridge Rural District Council in Zimbabwe, once remarked to one of us (DHMC), 'The problem is that cattle are mine but the wildlife is ours.'

In the early 1990s, the Nenyunga Ward, in the Gokwe District of Zimbabwe, still retained a rich wildlife population and tsetse fly had recently been eradicated with the result that cattle could be farmed in the ward. The option to maintain and develop a wildlife-based economy in the ward was still open. In a detailed study of the relative returns from wildlife and livestock in the ward, Murindagomo (1997) found that wildlife could be more profitable than cattle at a household level, provided cattle subsidies were removed and that appropriate institutions and policies governing access to wildlife resources and benefits were established. Since this did not happen, the wildlife resource has been eroded and potential benefits have evaporated.

In Botswana, the extent of resource management devolution to community-based organisations (CBOs) is severely constrained by centralised state and district controls (Martin, 2008; Rozemeijer, 2009). In order to engage in wildlife management CBOs are required to: (a) submit a land use and management plan to obtain a 15-year lease to use natural resources, (b) pay an annual land rental and a 'resource utilisation royalty', (c) pay 65 per cent of the income to a National Environmental Fund and provide regular reports on the use of the remainder of the income, (d) seek permission from the Land Authority to enter into a joint venture partnership, with a committee deciding on who the partner may be, and (e), participate in collecting natural resource socio-economic data.

As Martin (2008) convincingly argues, no cattle farmer would accept these conditions in order to graze his cattle and he goes on to question whether wildlife managers (whether individuals or communities) can seriously be asked to accept the same impositions.

In a study of policy issues in the Zambian component of the KAZA TFCA, Metcalfe (2006) drew attention to the sectoral and legal separation of forestry, fisheries and wildlife as a major constraint to effective natural resource management on the ground. These are difficulties that are common to most of the countries of Southern Africa where differing resource access rights and jurisdictions relate to timber, non-timber forest products, fish, water and wildlife. As Metcalfe (2006) effectively elaborated:

> The protected local and national forests are surrounded by communal land and provide valuable wildlife habitat, but no policy integrates land, forest and wildlife tenure or management. Communities depend on their traditional authorities for communal land management issues and on the forest and wildlife authorities for an unequal co-management relationship. Three separate legal, policy and institutional environments pertain and the common property design flaws in this arrangement mean high transaction costs, overlapping jurisdictions, and assure a 'tragedy of the commons' on the ground (Hardin, 1968).
> (Metcalfe 2006, p. 12)

Examples of potential returns and livelihood opportunities in TFCAs

Wildlife-based tourism

Two examples – one from the Kwando River in Caprivi in the heart of KAZA and one from the Kafue National Park and its surrounding GMAs – are presented here to illustrate the opportunities and returns that could be realised from wildlife-based enterprises for those living on the edge.

The Mudumu North Complex, which straddles the Kwando River, provides an example of current developments and opportunities for communities living on the edge. The complex covers 4,230 km^2 and comprises two national parks, Bwabwata to the west and Mudumu to the east of the Kwando River, the Caprivi Forest Area and three conservancies (Kwando, Mayuni and Mashi), which share boundaries with Bwabwata National Park along the Kwando and an emerging conservancy, Masidi. There is also an area of communal land that is not included within the conservancies. The area is well endowed with alluvial soils, carries a sizeable livestock population and a diverse and increasing wildlife population presently dominated by elephants. In a detailed examination of the wildlife potential of the area, Martin (2006) considered that the existing returns from wildlife were about one-third of those that could be realised if appropriate management of the area for wildlife, livestock and agriculture was introduced. The three state-protected areas cover 67 per cent of the complex, with communal land covering the remaining 37 per cent. The Kwando, Mayuni and Mashi conservancies cover 16 per cent of the complex. The study assumed that sport hunting in the complex would not be affected if half of the Kwando River frontage were reserved for non-consumptive eco-tourism. It also assumed that elephant numbers could be reduced to result in a more balanced wildlife population. Martin (2006) calculated the per hectare and capita returns from cattle and crops and from wildlife that could be expected for each of the six communal land units within the complex. The estimated per capita income from cattle and crops in the six land units varied between US$31 and US$106, and the per capita income from wildlife varied between US$13 and US$245 with the least returns from wildlife in the communal land section outside the conservancies. The potential overall returns within the complex from cattle and crops were estimated to be about US$5.2 per ha with an additional US$4.2 per ha from wildlife. As Martin (2006) concluded, 'If the optimum wildlife scenario can be realised it will add 50% or more to the present incomes of the communities in the Mudumu North Complex. This is a significant increment.'

In a study of existing returns to the Kafue National Park and surrounding GMAs, Martin (2007) noted that the tourism industry, in terms of both game viewing and hunting, depended on its wildlife populations. These were well below carrying capacity as a result of illegal hunting and the lack of meaningful returns to those surrounding the national park. The present management plans for the park focus on potential income for ZAWA, but it is important to see the national park and the surrounding GMAs in a wider context of opportunities that will benefit the national park, the GMAs and the people living within them, as well as the national fiscus. The present park budget does not meet 25 per cent of the funds required to realistically protect its wildlife under existing circumstances. The amounts required would be of the order of US$3.8 to US$5 million for the park and US$11 million for the GMAs, which cover more than twice the area of the park.

Martin (2007) prepared a detailed business model that involved increasing the number of lodges and camps in the park over a 12-year period, and the establishment of business partnerships between the concessionaires and the surrounding Community Resource Boards. If implemented over the 12-year period, the revenue generated could be increased from about US$2 million per annum in 2005/2006 levels to US$50 million per annum. Importantly, with the Community Resource Boards increasing their share holding over the 12-year period to a 49 per cent share in the business, there would be strong incentives for communities to protect their wildlife and the costs of running the park and protecting its wildlife resources would be small. The potential earnings and distribution of revenue from the Kafue National Park alone, if the plan is fully implemented over the 12-year period, are summarised in Figure 9.3. The return to the GMAs would be in the region of US$5–8 per ha, compared with current returns of less than US$0.5 per ha.

Combining wildlife and livestock

Historically, livestock keepers in east and Southern Africa were able to move their herds seasonally over large areas to take advantage of key grazing resources at appropriate times of the year. Before the advent of modern weapons, the overhunting of the eighteenth and nineteenth centuries, and the rinderpest pandemic in the late 1800s, livestock shared the range with herds of wildlife. The practice of using fences to restrict the movements of wildlife and livestock in Southern Africa is less than 100 years old. In Zimbabwe for instance, there was no national effort to construct fences to separate cattle and wildlife until the 1950s (Foggin, 2010), even after the role played by buffalo in the maintenance and spread of important diseases to cattle had been demonstrated, such as the fatal corridor disease and the highly contagious Foot and Mouth Disease. However, the control of trypanosomiasis transmitted by tsetse flies, which normally feed on various wild species, led to the erection of fences and the elimination of buffalo, antelope and wild suids from designated areas in the Southeast Lowveld during the 1960s (Condy, 1979; Child and Riney, 1987).

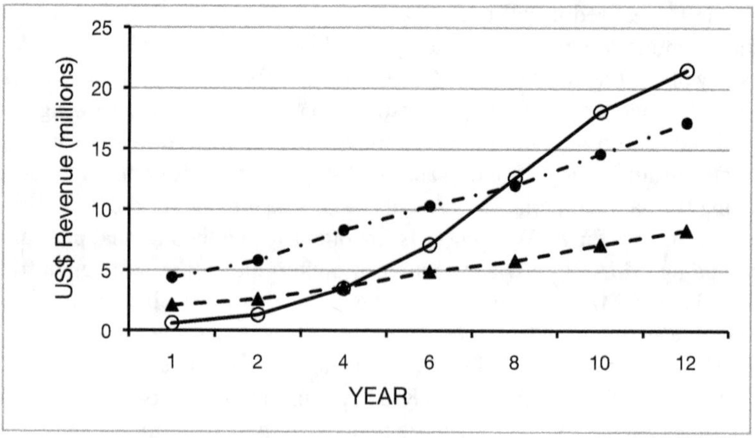

Figure 9.3 Projected returns from joint venture tourism operations in and surrounding Kafue National Park starting in year 1 with current (2007) income. Revenues to Community Resource Boards (open circles), Hunting and Non-Hunting Tour Operators (closed circles) and returns to government (taxes and fees to Zambian Wildlife Authority) (closed triangles).

Source: Martin (2007).

However, there are a number of ecological, economic and even political advantages to production systems that integrate livestock and wildlife (Jansen *et al.*, 1992; Cumming 1995). The association on the same pastures of several species of herbivores with different diets allows a better use of the vegetation and helps to reduce bush encroachment (Taylor and Walker, 1978). Controlled grazing of cattle herds can be used as a management tool to maximise the use of pastures, while wild species usually move freely between cattle paddocks and adapt their diet and habitat preferences accordingly (Fritz *et al.*, 1996). In addition, from an economic point of view, combining cattle production and wildlife-based tourism provides greater income, better margins and increased resilience to the risks associated with uncertain climatic conditions, fluctuating beef markets and tourist activity. On the other hand, it has been argued that in some situations (e.g in Kenya) there are significant opportunity costs, resulting from diseases and lack of access to export markets, associated with land carrying both wildlife and livestock compared to those without wildlife (Kock, 2010). In the context of strong pressures on land and land reform, it is politically imperative to demonstrate that a system is maximising land productivity, and that integrated wildlife–livestock production systems have an advantage.

Communities and government increasingly appreciate the advantages of running multi-species systems in the community conservancies in Namibia (Weaver and Skyer, 2005; NACSO, 2010). In Zimbabwe, the Agriculture and Rural Development Agency (ARDA) managed Battlefields Estates during the 1990s as a mixed wildlife–livestock production system (de Garine-Wichatitsky, 1999). A comprehensive study of 140 commercial ranches carrying livestock, wildlife and livestock alone revealed the economic and financial benefits of multi-species systems (Jansen *et al.*, 1992). Effective, profitable and resilient multi-species systems that include the full range of wildlife and the 'big five' are developing in the Laikipia region of Kenya. The Ol Pejeta Conservancy provides a clear example where a pedigree herd of *c.*6,000 Boran cattle is successfully managed alongside the full range of wildlife species, including large predators.[2]

Opportunities thus exist in some large TFCAs, such as the GLTFCA and KAZA, and in large semi-arid landscapes in Southern Africa, to manage mobile herds of livestock alongside wildlife populations in ways that could improve returns to communities and result in improved rangeland management and productivity (Cumming, 1995; Cumming and Bond, 1991). To some extent, such agreements have been initiated between managers of protected areas in Zimbabwe and neighbouring communities, who have been granted permission to graze their cattle across the edge during periods of extreme fodder shortage. For instance, managers of GNP granted this permission during the droughts of 1992 and 2005, and there is a tolerance during the dry season for incursions of cattle inside the Sikumi Forest situated on the northern boundary of Hwange National Park (HNP).

Technical developments to promote greater synergy between wildlife/livestock production systems

Present concerns with animal diseases (see Chapters 3 and 8), particularly FMD, and the associated fences and restrictions on export trade in animal products in large parts of the region could be mitigated, if not overcome, by technical or regulatory innovations (e.g. Scoones *et al.*, 2010). Improved vaccines will serve this objective in several ways: they will provide better protection of livestock herds in contact with wild reservoir hosts, safer and cheaper vaccines (e.g. Theileriosis; Morzaria *et al.*, 2000; Latif *et al.*, 2002), and the use of recombinant vaccines associated with improved diagnostic tests, will allow the discrimination between vaccinated and infected animals.

Finding ways for livestock owners living on the edge to market the products of their livestock is of paramount importance in the context of expanding TFCAs in the region (Thomson, 2008). One possible solution is the introduction of commodity-based trade, which argues that some products from healthy animals (e.g. de-boned beef; Thomson *et al.*, 2009) can be processed in a way that guarantees a minimal risk of transmitting diseases, even if the animals are living in an area that has diseases of economic importance such as FMD. This principle is gradually gaining acceptance by national authorities in the region and by international donor agencies, but has still to be accepted by the OIE, the agency responsible for international trade regulations for animal products.

Using fences can reduce human–wildlife conflict. Numerous different types of fences have been designed, and fencing technology and the use of solar power has advanced considerably over the last decade (see review, Ferguson and Hanks, 2010). However, the costs of maintaining fences can be high and beyond the capacity of rural communities. The Masoka community in the Zambezi Valley provided a workable example when a perimeter fence was erected around the village instead of attempting to fence in, or fence off, a much larger wildlife area.

Another recent development is the use of mobile predator-proof kraals or bomas in which livestock can be confined overnight. The kraals are assembled from segments that can be constructed of woven branches or from steel frames and wire netting. The walls need to be about 3 metres in height and the area of the kraal such that the animals inside are sufficiently closely packed to prevent a stampede should a predator approach the kraal. Grass cover and grazing intensity by livestock can be managed more selectively by moving herds to improve pastures to benefit both wildlife and livestock, while the dung deposited in such kraals can provide durable nutrient hotspots in the range (e.g. Young *et al.*, 1995).

Partnerships and joint ventures

Partnerships in tourism ventures have been successfully established between tour companies and communities in Namibia and between companies and CBOs in Botswana (Rozemeijer, 2009). In South Africa, community property associations are able to establish lease arrangements with the private sector as, for example, have the Makuleke community. In Zimbabwe, tourism and trophy hunting lease arrangements are limited to RDCs because villages and wards do not have a legal persona. The development of local level resource management trusts offers a means for communities to enter into joint ventures but development of these is still constrained by RDC involvement and control.

More recently, the notion of private–public–community partnerships has been proposed in which state entities, such as national parks, together with communities, become a part of a wider commercial enterprise intended to stimulate rural development and tangible benefits to rural communities. However, little material progress appears to have been made in implementing such arrangements in the TFCAs in the region.

The potential for wildlife-based partnerships and joint ventures to make significant contributions to rural livelihoods, as evidenced by the Namibian Community Conservancies, the Kafue example, and other examples, is considerable but constrained by current policy frameworks and mistrust between potential partners at several levels.

Carbon trading and climate change issues

Carbon markets, linked to the conservation of forests and woodlands as a means of mitigating the emission of greenhouse gases and global warming, offer a potential source of revenue to those living on the edge in TFCAs. The initiative to Reduce Emissions from Deforestation and Forest Degradation in developing countries (REDD) is gaining momentum (Anonymous, 2008; Phelps *et al.*, 2010; Sandker *et al.*, 2010). TFCAs in higher rainfall areas (e.g. Niassa-Selous and parts of the KAZA TFCA), where there are substantial areas of miombo woodland, stand to benefit from the REDD programme. There are, however, major hurdles in terms of legal and governance issues to be resolved in ensuring that revenues reach the rural poor, and in monitoring compliance in forest conservation targets (Chhatre and Agrawal, 2009; Corbera *et al.*, 2010; Ghazoul *et al.*, 2010; Streck, 2010). Bond *et al.* (2010) provide a thorough analysis of the prospects for, and the issues involved in, pro-poor development of carbon benefits to villagers within the miombo woodlands of east and Southern Africa. The extent to which forest or woodland conservation will also result in the conservation of other components of biodiversity, such as vertebrates, will also require examination (Venter *et al.*, 2009).

Scientific, technical and cultural exchanges

Scientific, technical and cultural exchanges across international boundaries form an important and necessary component of harmonising resource management, disease control, commercial activities and values regarding resource use and conservation within TFCAs. Scientific and technical exchanges and cooperation preceded the formation of the Kgalagadi Transfrontier Park by many years (Schoon, 2008). Scientific exchanges between Mozambique, South Africa and Zimbabwe in the GLTFCA have been particularly strong over the last eight years through the AHEAD (Animal and Human Health for the Environment and Development)-GLTFCA programme and direct contacts between government agencies. The AHEAD-GLTFCA Working Group has convened multidisciplinary meetings that each involve scientists from the three countries and beyond, the proceedings of which are available at the AHEAD website.[3]

Technical cooperation across boundaries has involved coordinated or joint aerial surveys of large mammals, particularly of elephant, e.g. between Zimbabwe and Botswana on the boundary of HNP, between Zimbabwe and Zambia across the Zambezi River in the Mana Pools-Lower Zambezi TFCA. It has involved capture and translocation of animals as between the KNP and the Limpopo National Park (LNP). The free movement of scientists across borders within a TFCA when engaged in cross-boundary research or management (e.g. tracking tagged animals), however, appears to have been established only for the Kgalagadi Transfrontier Park.

Transboundary natural resource management at a local level that involves local authorities and people living on international boundaries is still weakly developed, if at all. In the GLTFCA, recent initiatives have been taken to establish a transboundary natural resource management forum comprising members from local government and traditional leaders in each of the districts adjoining the central point of the GLTFCA where the boundaries of Mozambique, South Africa and Zimbabwe meet (Cumming *et al.*, 2009). A local transboundary community resource management forum has also been established between Namibia and Zambia in the KAZA TFCA.

For those TFCAs that have treaties or memoranda of understanding between the countries involved, provision is made to establish joint management boards and technical committees covering the various facets of TFCA management such as veterinary and conservation issues, tourism, and immigration and security. These committees involve central government officials from the participating countries and may call on additional expertise when needed.

The rhetoric surrounding the creation of TFCAs has often promoted the notion of facilitating the re-establishment of cultural ties and links across boundaries, however, the requirements of immigration authorities, and security and nationalist concerns (Metcalfe and Kupe, 2008), mean that cross-border movement remains difficult. In most cases the people living on either side of international boundaries within TFCAs belong to the same tribal group and close familial and cultural ties exist. These are particularly strong in relation to issues such as land fertility rituals and formalised rituals of transition to adulthood. In some cases cross-border movement has been facilitated (Ramutsindela, 2007).

Conditions for moving from opportunities to implementation

Issues of empowerment and natural resource governance lie at the centre of moving from opportunities at the edge within TFCAs to their realisation. The division of responsibilities for natural resources into separate sectors, as noted by Metcalfe (2006), empowers the state. The result is that these resources (forests, fish, wildlife and water) are perceived as belonging to the state and, in the inevitable absence of effective controls, become open access commons susceptible to collapse. So what are the conditions and strategies that are needed to move from unrealised opportunities to their effective implementation?

TFCAs are complex social-ecological systems (SES) and, as Ostrom (2007, 2009) has argued, there is a need to move beyond panaceas and the expectation that simple recipes will resolve the problems of managing the commons. Self-organisation of resource users is central as are the conditions (policies and institutions) that allow this to happen. The management of natural resources across boundaries, whether between land tenure regimes or across international borders, requires adaptive co-management regimes, for which essential requirements are appropriate policy and legal frameworks that both facilitate and allow co-management regimes to develop (Ruitenbeek and Cartier, 2001).

With particular reference to analysing and managing common property regimes, Ostrom (2007, 2009) outlined a multi-tier framework comprising the resource system, the governance system and resource units and users embedded in a wider social, economic and political system and related ecosystems. She further identified 10 core variables associated with self-organisation and achieving sustainable SESs, namely, (a) size of the resource system, (b) productivity of the system, (c) predictability of system dynamics, (d) resource unit mobility, (e) collective choice rules, (f) number of users (and transaction costs), (g) leadership/entrepreneurship, (h) norms and social capital (trust and reciprocity), (i) knowledge of SES/mental models and (j) importance of the resource.

For sustainable natural resource management in TFCAs to function, and for effective self-organisation to begin, the social, economic and political settings generally need to be conducive. It is in this context that there is an urgent need for policy reform. The reforms most urgently needed relate to devolution of secure and tradable resource access rights, and empowerment at local levels that results in the distribution of economic power and the ability to hold local and central governments accountable (Ostrom and Janssen, 2002; Quadir, 2010). Centralised top-down solutions in the form of aid and large rural development projects have seldom worked (Collier, 2007; Moyo, 2009) and are unlikely to work, in the context of rural development in the TFCAs within the region.

Strategies for reform and empowerment of local resource managers

Associated with the needs for policy reform and empowerment is the need to revisit the notion that communal and other land outside of protected areas within TFCAs can be sustained on wildlife-based production systems alone. Cumming (2005) outlined four strategic approaches for the Southeast Lowveld (SEL) of Zimbabwe and the GLTFCA that could be more widely applicable.

Applying the following four strategies would go some way towards realising the opportunities for those living on the edge within TFCAs in Southern Africa.

1. To place a premium on, and invest in, higher valued land uses and diversification. In most TFCAs there are areas of high agriculture potential, such as areas of irrigable soil that would merit development. Developing potential intensive production areas, in concert with appropriate livestock development, would go some way towards alleviating food shortages and unemployment. Associated investments in infrastructure to facilitate marketing of goods and services would be necessary.
2. Decouple wealth creation from net above ground primary production. Because primary production in arid and semi-arid areas is greatly limited by rainfall (and in many high rainfall areas production is limited by nutrients) the more that wealth creation can be decoupled from a direct reliance on primary and secondary production, the less susceptible it will be to annual and seasonal fluctuations in rainfall. One means of achieving this end is to develop high-valued tourism ventures where the value is derived from services instead of from crop and meat production. Payments for ecosystem services offer an additional and growing opportunity.
3. Match land use and ecological process scales. Livestock and wildlife production systems generally require large areas over which to exploit temporal and spatial variations in the availability of key resources. Fragmentation of large landscapes by fencing and inappropriate land tenure and systems of resource access rights militate against adaptive strategies that may be more productive and sustainable. The development of large-scale wildlife conservancies involving the effective amalgamation of former cattle ranches into large-scale wildlife tourism areas is a case in point (e.g. du Toit, 1992).

4. Develop legal and policy frameworks that enable local level innovation and adaptability in resource access rights and management strategies, and in production and marketing. Current centralised prescriptions over land use, tenure and resource access rights and marketing effectively stifle innovation and the development of adaptive co-management regimes (e.g. Ruitenbeek and Cartier, 2001) at larger scales and across land tenure categories. It is suggested that releasing the innovative capacities of farmers, resource managers and communities may go a long way towards solving the food and environmental security problems of those living on the edge.

Concluding comment

An overview of the resource base in terms of soils and rainfall in the 13 TFCAs being developed in Southern Africa reveals that most TFCAs are situated on marginal land in terms of productivity, infrastructure and services. Were it not so, these areas would long since have been settled and used for agriculture. As a result TFCAs are unlikely to be able to support dense human populations. In many areas, the present populations are subsidised by external, off-farm remittances and periodic injections of food aid. While some measure of agricultural development will help in many cases, the highest valued land use is most likely to be wildlife-based tourism and a full diversification of natural resource management and resource use options. The development of Namibian communal conservancies provides very clear support for such an approach. This is reinforced by the several examples that we have provided of potential returns from a wide range of natural resource-based enterprises that include non-consumptive tourism, safari hunting, the harvesting of meat, crafts, non-timber forest products and the provision of ecosystem services such as water and carbon sequestration. The overriding constraints in most of Southern Africa to improved natural resource-based livelihoods for those living on the edge are those related to institutions (the rules of the game) and to centralised national policy and legal frameworks. As Metcalfe has so aptly stated:

> Provided with legal, enterprise and organization support local communities may just be capable of creating upward pressure to secure their rights. This would seem a more promising approach than waiting patiently for rights to be given. Optimistically, transfrontier initiatives may provide a new policy forum to make their 'voices' heard.
>
> (Metcalfe, 2006, p. 70)

Up to now the planning, design and management of TFCAs has rested at central government level. The views of the de facto resource managers living within the matrix of communal lands within TFCAs have seldom been heard if at all. A key to unlocking those voices and their energies in developing TFCAs is to bring them into the fold.

Acknowledgements

Harry Biggs, Graeme Cumming, Brian Jones and Russell Taylor read a penultimate draft of this chapter and we thank them for their critical and constructive reviews. We also thank Meg Cumming for copyediting and commenting on successive drafts.

Notes

1 For the purposes of this chapter, Southern Africa is taken to be Angola, Zambia and Tanzania and the countries to the south of these, although we draw on some examples from Kenya in East Africa.
2 www.olpejetaconservancy.org.
3 www.wcs-ahead.org/workinggrps_limpopo.html.

References

Anonymous (2008) *UN Collaborative Programme on Reducing Emissions from Deforestation and Forest Degradation in Developing Countries (UN-REDD)*, FAO, UNDP, UNEP Framework Document, 20 June 2008, Rome

Anstey, S. (2008) 'Institutional Change and Community Based Natural Resource Management in Northern Mozambique', Ph.D. thesis, Centre for Applied Social Sciences, University of Zimbabwe, Harare

Bandyopadhyay, S. and Tembo, G. (2009) 'Household welfare and natural resource management around national parks in Zambia', Policy Research Working Paper no. 4932, Environment Department, The World Bank Washington DC

Bennett, B. (2006) 'Natural Products: the New Engine for African Trade growth', RTFP, IUCN SA, PhytoTrade Africa. Available at: www.earthmind.net/biotrade/docs/sa/iucn-natpro-trade.pdf (accessed 14 May 2011)

Bond, I. and Cumming D.H.M. (2006) 'Wildlife research and development', in M. Rukuni, P. Tawonezi, C. Eicher, M. Munyuki-Hungwe and P. Matondi (eds), *Zimbabwe's Agricultural Revolution Revisited*, 2nd ed., University of Zimbabwe Publications, Harare

Bond, I., Chambwera, M., Jones, B., Chundama, M. and Nhantumbo, I. (2010) 'REDD+ in dryland forests: issues and prospects for pro-poor REDD in the miombo woodlands of Southern Africa', *Natural Resource Issues*, no. 21, IIED, London

Chennells, R. (2001) 'The Khomani San of South Africa', in *Indigenous Peoples and Protected Areas in Africa: From Principles to Practice*, Forest People's Programme, Case Study 9. Available at: www.forestpeoples.org/es/resources/results/taxonomy:23.79 (accessed 13 August 2010)

Chhatre, A. and Agrawal, A. (2009) 'Trade-offs and synergies between carbon storage and livelihood benefits from forest commons', *Proceedings of the National Academy of Sciences*, vol. 106, pp. 17667–17670

Child, B.A. (ed.) (1995) *Parks in Transition: Biodiversity, Rural Development and the Bottom Line*, Earthscan, London

Child, G.F.T. and Riney, T. (1987) 'Tsetse control hunting in Zimbabwe: 1919–1958', *Zambezia*, vol.14, pp. 11–71

Cirelli, M.T. and Morgera, E. (2010) 'Wildlife Law in the Southern African Development Community', CIC Technical Series Publication No. 9, Joint Publication of FAO and CIC, Budapest

Collier, P. (2007) *The Bottom Billion: Why the Poorest Countries Are Failing and What Can Be Done About It*, Oxford University Press, Oxford

Condy, J.B. (1979) 'A history of foot and mouth disease in Rhodesia', *Rhodesian Veterinary Journal*, vol. 10, pp. 2–10

Corbera, E., Estrada, M. and Brown, K. (2010) 'Reducing greenhouse gas emissions from deforestation and forest degradation in developing countries: revisiting the assumptions', *Climatic Change*, vol. 100, pp. 355–388

Cumming, D.H.M. (1995) 'Are multispecies systems a viable landuse option for Southern African savannas?', in R.R. Hofmann and H.J. Schwartz (eds), *International Symposium on Wild and Domestic Ruminants in Extensive Land Use Systems*, Okolgische Hefte der Landwirtschaftlich-Gartnerischen Fakulteit Heft 2, Humbolt-Universitat zu Berlin, Berlin

Cumming, D.H.M. (2005) 'Wildlife, livestock and food security in the South-East Lowveld of Zimbabwe', in S.A. Osofsky, S. Cleaveland, W.B. Karesh, M.D. Kock, P.J. Nyhus, L. Starr and A. Yang (eds), *Conservation and Development Interventions at the Wildlife/Livestock Interface: Implications for Wildlife, Livestock, and Human Health*. IUCN Occasional Paper no. 30, Gland, Switzerland and Cambridge, UK, pp. 41–45

Cumming, D.H.M., Bocchino, C. and Caroli, P. (2009) 'Exploring options for a regional forum for transboundary natural resource management in the Limpopo Corridor, Record of the Tshipise Meeting, November, 2009', CESVI, Cape Town

Cumming, D.H.M. and Bond, I. (1991) 'Animal production in Southern Africa: present practice and opportunities for peasant farmers in arid lands', WWF Multispecies Animal Productions Systems Project, Project Paper No. 22, Harare

Cumming, G.S., Cumming, D.H.M. and Redman, C.L. (2006) 'Scale mismatches in social-ecological systems: causes, consequences, and solutions', *Ecology and Society*, vol. 11, no. 1, p. 14. Available at: www.ecologyandsociety.org/vol11/iss1/art14/

Denkler, J. L. (2009) 'Community based natural resource management: power, isolation, and development in rural Botswana', MA thesis, University of Florida, Gainsville

Ferguson, K. and Hanks, J. (eds) (2010) *Fencing Impacts: A review of the Environmental, Social and Economic Impacts of Game and Veterinary Fencing in Africa with Particular Reference to the Great Limpopo and Kavango-Zambezi Transfrontier Conservation Areas*, Mammal Research Institute, University of Pretoria, Pretoria. Available at: www.wcs- ahead.org/gltfca_grants/grants.html

Fernandez, A. (2009) 'Wildlife conservation in Zambia: impact of game management areas on household welfare', MSc thesis, Michigan State University, Michigan

Foggin, C. (2010) 'Zimbabwe's livestock disease control policy in relation to fencing', in K. Ferguson and J. Hanks (eds), *Fencing Impacts: A Review of the Environmental, Social and Economic Impacts of Game and Veterinary Fencing in Africa with Particular Reference to the Great Limpopo and Kavango-Zambezi Transfrontier Conservation Areas*, Mammal Research Institute, University of Pretoria, Pretoria

Fritz, H., Garine-Wichatitsky, M. de and Letessier, G. (1996) 'Habitat use by sympatric wild and domestic herbivores in an African savanna woodland: the influence of cattle on spatial behaviour', *Journal of Applied Ecology*, vol. 33, pp. 589–598

Garine-Wichatitsky, M. de (1999) 'Ecologie des interactions hôtes/vecteurs: analyse du système tiques/ongulés sauvages et domestiques en zone tropicale', Ph.D. thesis, Université des Sciences et Techniques du Languedoc, Montpellier II, Montpellier

Ghazoul, J., Butler, R.A., Mateo-Vega, J. and Koh, L.P. (2010) 'REDD: a reckoning of environment and development implications', *Trends in Ecology and Evolution*, vol. 25, pp. 396–402

Grossman, D. and Holden P. (2009) 'Towards transformation: contractual national parks in South Africa', in H. Suich, B. Child and A. Spenceley (eds), *Evolution and Innovation in Wildlife Conservation: Parks and Game Ranches to Transfrontier Conservation Areas*, Earthscan, London

Hardin, G. (1968) 'The tragedy of the commons', *Science*, vol. 162, no. 3859, pp. 1243–1248

Jansen, D., Bond, I. and Child, B. (1992) 'Cattle, wildlife, both or neither: a summary of survey results for commercial ranches in Zimbabwe', WWF Multispecies Animal Production Systems Project Paper no. 30, WWF, Harare

Jones, B. (2008) 'Legislation and polices relating to protected areas, wildlife conservation, and community rights to natural resources in countries being partner in the Kavango Zambezi Transfrontier Conservation Area', Consultant's Report to Conservation International, Cape Town

Jones, B. and Weaver, L.C. (2009) 'CBNRM in Namibia: growth, trends, lessons and constraints', in H. Suich, B. Child and A. Spenceley (eds), *Evolution and Innovation in Wildlife Conservation: Parks and Game Ranches to Transfrontier Conservation Areas*, Earthscan, London

Jones, B.T.B. (2009) 'A study of the policy and legal framework governing transboundary natural resource management in the Great Limpopo Transfrontier Conservation Area', Consultant's Report to Limpopo Transboundary Programme, CESVI, Cape Town

Kock, R.A. (2010) 'The newly proposed Laikipia disease control fence in Kenya', in K. Ferguson and J. Hanks (eds), *Fencing Impacts: A Review of the Environmental, Social and Economic Impacts of Game and Veterinary Fencing in Africa with Particular Reference to the Great Limpopo and Kavango-Zambezi Transfrontier Conservation Areas*, Mammal Research Institute, University of Pretoria, Pretoria

Latif, A.A., Hove, T., Kanhal, G.K. and Masaka, S. (2002) 'Buffalo-associated *Theileria parva*: the risk to cattle of buffalo translocation into the highveld of Zimbabwe', *Annals of the New York Academy of Science*, vol. 969, pp. 275–279

Le Breton, G. and Nemarundwe, N. (2009) 'Linking Southern African Producers of Non-timber Forest Products to Global Markets', XIII World Forestry Congress, Buenos Aires, Argentina, 18–23 October 2009

Lewis, D.M. and Alpert, P. (1997) 'Trophy hunting and wildlife conservation in Zambia', *Conservation Biology*, vol. 11, pp. 59–68.

Martin, R.B. (2006) 'The Mudumu North Complex: wildlife co-management in the Kwando Area', a study for the Ministry of Environment and Tourism and the Management Committee of the Mudumu North Complex, Windhoek

Martin, R.B. (2007) 'Potential of tourism-based business of Kafue National Park and associated GMAs', Unpublished Consultant's Report

Martin, R.B. (2008) 'Development of strategic options to improve the performance of the Botswana Department of Wildlife and National Parks: Final Report', Atos Origin, Belgium, and Department of Wildlife and National Parks, Ministry of Environment, Wildlife and Tourism, Botswana

Metcalfe, S.C. (2006) 'Communal land reform in Zambia: governance, livelihood and conservation', MPhil thesis, Programme for Land and Agrarian Studies (PLAAS), Faculty of Economic and Management Sciences, University of the Western Cape, Cape Town

Metcalfe, S. and Kupe, T. (2008) '"Your elephant on our land." The struggle to manage wildlife mobility on Zambian Communal Land in the Kavango–Zambezi Transfrontier Conservation Area', *Journal of Environment and Development*, vol. 17, pp. 99–117

Morzaria, S.P., Nene, V., Bishop, R. and Musoke, A.J. (2000) 'Vaccines against *Theileria parva*', *Annals of the New York Academy of Science*, 196, pp. 464–473

Moyo, D. (2009) '*Dead Aid: Why Aid is Not Working and How There is Another Way For Africa*', Allen Lane, London

Murindagomo, F. (1997) 'Wildlife, cattle and comparative advantage in semi-arid communal lands and implications for agropastoral options and government policy: a case study in the Sebungwe Region, Zimbabwe', Ph.D. thesis, Department of Agricultural Economics and Extension, University of Zimbabwe, Harare

Murphree, M.W. (1995) 'The lesson from Mahenye: rural poverty, democracy and wildlife conservation', *Wildlife and Development Series*, No1. IIED, London

NACSO (2010) *Namibia's Communal Conservancies: A Review of Progress in 2010*, Namibian Association of CBNRM Support Organisations. Available at: www.nacso.org.na/SOC_2009/index.php (accessed 12 May 2011)

Nelson, F. and Agrawal, A. (2008) 'Patronage or participation? Community-based natural resource management reform in Sub-Saharan Africa', *Development and Change*, vol. 39, pp. 557–585

Nhantumbo, I. and Anstey, S. (2009) 'CBNRM in Mozambique: the challenges of sustainability', in H. Suich, B. Child and A. Spenceley (eds), *Evolution and Innovation in Wildlife Conservation: Parks and Game Ranches to Transfrontier Conservation Areas*, Earthscan, London

Norfolk, S. and Tanner, C. (2007) 'Improving tenure security for the rural poor: Mozambique country case study', Legal Empowerment for the Poor (LEP), Working Paper #5, FAO, Rome

Ostrom, E. (2007) 'A diagnostic approach for going beyond panaceas', *Proceedings of the National Academy of Sciences*, vol. 104, pp. 15181–15187

Ostrom, E. (2009) 'A general framework for analyzing sustainability of social-ecological systems', *Science*, vol. 352, pp. 419–422

Ostrom, E. and Janssen, M.A. (2002) 'Beliefs, multi-level governance, and development', Bloomington, Indiana. Draft, 28/8/02, Prepared for the 2002 Annual Meeting of the American Political Science Association, Boston, Massachusetts, 29 August–1 September, 2002

Peterson, J.H. (1991) 'A proto-CAMPFIRE initiative in Mahenye Ward, Chipinge District: development of a wildlife programme in response to community needs', Unpublished Report, Centre for Applied Social Science, University of Zimbabwe, Harare

Phelps, J., Guerrero, M., Dalabajan, D., Young, B. and Webb, E. (2010) 'What makes a "REDD" country?', *Global Environmental Change*, vol. 20, pp. 322–332

Quadir, I. (2010) 'Empowerment is the key', *Nature*, vol. 465, pp. 550–551

Ramutsindela, M. (2007) *Transfrontier Conservation in Africa: At the Confluence of Capital, Politics and Nature*, CABI Publishing, Wallingford

Reid, H., Fig, D., Magome, H. and Leader-Williams, N. (2010) 'Co-management of contractual parks in South Africa: lessons from Australia', *Conservation and Society*, vol. 2, no. 2, pp. 377–409

Rihoy, E., Chirozva, C. and Anstey, S. (2004) '"People are not happy" – speaking up for adaptive natural resource governance in Mahenye', Occasional Paper no. 31, Land and Agrarian Reform in Southern Africa Occasional Paper Series, University of the Western Cape, Cape Town

Robins, S. (2001) '"Bushmen" and double vision: the Khomani San land claims and the cultural politics of "community" and development in the Kalahari', *Journal of South African Studies*, vol. 27, pp. 833–853

Rozemeijer, N. (2009) 'CBNRM in Botswana', in H. Suich, B. Child and A. Spenceley (eds), *Evolution and Innovation in Wildlife Conservation: Parks and Game Ranches to Transfrontier Conservation Areas*, Earthscan, London

Ruitenbeek, J. and Cartier, C. (2001) *The Invisible Wand: Adaptive Co-Management as an Emergent Strategy in Complex Bio-Economic Systems*, Occasional Paper 34, 1–47, CIFOR, Bogor

Sandker, M., Nyame, S.K., Förster, J., Collier, N., Shepherd, G., Yeboah, D., Blas, D.E. de, Machwitz, M., Vaatainen, S., Garedew, E., Etoga, G., Ehringhaus, C., Anati, J., Quarm, O.D.K. and Campbell, B.M. (2010), 'REDD payments as incentive for reducing forest loss', *Conservation Letters*, vol. 3, pp. 114–121

Schoon, M.L. (2008) 'Building robustness to disturbance: governance in Southern African Peace Parks', Ph.D. thesis, Indiana University, Bloomington

Scoones, I., Bishi, A., Mapitse, N., Moerane, R., Penrith, M-L., Sibanda, R., Thomson, G. and Wolmer, W. (2010) 'Foot-and-mouth disease and market access: challenges for the beef industry in Southern Africa', *Pastoralism*, vol. 1, no. 2, pp. 135–164

Stack, J., Dorward, A., Gondo, T., Frost, P., Taylor, F. and Kurebgaseka, N. (2003) 'Mopane worm utilization and rural livelihoods in Southern Africa', Paper presented at the International Conference on Rural Livelihoods, Forests and Biodiversity, 19–23 May 2003, Bonn

Steenkamp, C.I. and Grossman, D. (2001) 'People and parks: Cracks in the paradigm', IUCN South Africa Policy Think Tank 10, IUCN South Africa Country Office, Pretoria

Streck, C. (2010) 'Reducing emissions from deforestation and forest degradation: national implementation of REDD schemes', *Climatic Change*, vol. 100, pp. 389–394

Taylor, R.D. and Walker, B.H. (1978) 'Comparisons of vegetation use and herbivore biomass on a Rhodesian game and cattle ranch', *Journal of Applied Ecology*, vol. 15, pp. 565–581

Thomson, G.R. (2008) 'A short overview of regional positions on foot-and-mouth disease control in Southern Africa', in *Transboundary Animal Disease and Market Access: Future Options for the Beef Industry in Southern Africa*, Institute of Development Studies, Working Paper 2, Brighton

Thomson, G.R., Leyland, T.J. and Donaldson, A.I. (2009) 'De-boned beef – an example of a commodity for which specific standards could be developed to ensure an appropriate level of protection for international trade', *Transboundary and Emerging Diseases*, vol. 56, pp. 9–17

du Toit, R.F. (1992) 'Large-scale wildlife conservancies in Zimbabwe: opportunities for commercial conservation of endangered species', in W. Van Hoven, H. Ebedes and A. Conroy (eds), *Wildlife Ranching. A Celebration of Diversity*, South African Game Organization, Pretoria, pp. 295–300

Venter, O., Laurance, W.F., Iwamura, T., Wilson, K.A., Fuller, R.A. and Possingham, H.P. (2009) 'Harnessing carbon payments to protect biodiversity', *Science*, vol. 326, p. 1368

Weaver, L.C. and Skyer, P. (2005) 'Conservancies: integrating wildlife land-use options into the livelihood, development and conservation strategies of Namibian communities', in S.A. Osofsky, S. Cleveland, W.B. Karesh, M.D. Kock, P.J. Nyhus, L. Starr and A. Yang (eds), *Conservation and Development Interventions at the Wildlife/Livestock Interface: Implications for Wildlife, Livestock and Human Health*, IUCN, Gland and Cambridge

Young, T.P., Partridge, N. and Macrae, A. (1995) 'Long-term glades in acacia bushland and their edge effects in Laikipia, Kenya', *Ecological Applications*, vol. 5, pp. 97–108

10 Whither TFCAs and people on the edge in Southern Africa?

David H.M. Cumming, Jens A. Andersson, Michel de Garine-Wichatitsky, Vupenyu Dzingirai and Ken E. Giller

Introduction

Transfrontier conservation areas (TFCAs) are the latest chapter in the development of approaches to nature conservation, connecting conservation and development in inhabited conservation landscapes. The TFCAs of Southern Africa are essentially centrally driven enterprises that are planned to cover approximately 723,000 km^2, or some 10.5 per cent of the region. The largest TFCAs, such as KAZA, Niassa-Selous and the GLTFCA, are larger than some countries in Southern Africa. They encompass extensive areas of land under customary 'communal' tenure that are the home and source of livelihood for several million people. TFCAs bring a new approach in recognising that issues of conservation transcend national boundaries that call for common approaches. But TFCAs represent a new level of management sandwiched between influences at the global/international level and the national level (Figure 10.1). Here we reflect on the analyses presented in the previous chapters and explore what the future may hold for people living in these areas, on the edge of protected wildlife areas of global importance.

Reflections on previous chapters

An exploration of the literature in Chapter 2 revealed that the formation of TFCAs is essentially a politically driven process, and therefore their conservation and development value is often limited. The ecological discourse adopted in their promotion tends to ignore the existence of those living within TFCAs. Meanwhile, development becomes a 'deliverable' to an abstract category of 'local beneficiaries' or 'the community'. Rather than active participants and developers of TFCAs, local people remain a category of ill-understood resource users: 'flat characters', or stereotyped caricatures, with no history.

Over the last century, the boundaries of conservation areas have been far from static and the 'edge' has been a moving target. Chapter 3 highlighted the range of drivers responsible for creating and changing boundaries in areas now included in TFCAs. An expanding, settler-dominated, capitalist economy was a major force underpinning boundary formation processes, most notably land tenure divisions. Capitalist expansion also contributed to uncontrolled hunting which, together with the 1896 rinderpest pandemic, decimated previously abundant game populations. In the early twentieth century, marginal, often sparsely populated, lands on national boundaries were seen as areas to be set aside as game reserves. Animal disease control issues sometimes informed their formation and dominated the construction of fences across the region to control the movements of wildlife and livestock and the spread of diseases such as Foot and Mouth Disease (FMD). Further capitalist expansion and rapid population growth shifted disease infestation boundaries in the mid-twentieth century, and increased pressure on natural resources on the edge of protected areas. Wildlife utilisation and eco-tourism became increasingly important economic ventures, but generally meant shifts in land use and new fences, rather than new boundaries. The alignment and re-alignment of fences, often for differing purposes, has thus played a key role in defining the location and nature of the edge.

The human population in the region has increased 20-fold over the last century. Chapter 4 examined the drivers of population change on the edge. Human populations within TFCAs have increased drastically, although in some localities much more than others. The result of in-migration, including forced displacements from elsewhere, is a huge range in livelihood pursuits and a high dependence on natural resources. Ecological factors, particularly the distribution of soils and water, still shape livelihoods for people living in TFCAs because of their reliance on natural resources and limited involvement in conservation-based income-generating activities. In addition, remittances play an important role in resource accumulation on the edge and are central for many households. A key point that emerged is that drivers other than conservation policies have probably played a greater role in shaping developments on the edge. Conservation has not attracted additional people, but people have moved into these areas because of lack of opportunities elsewhere.

Issues relating to ethnic diversity and its implications for managing resources in TFCAs were taken up in Chapter 5. The various senses in which 'ethnicity' is defined and used were examined, as were the ways in which it can be invented and used to mobilise or capture resources. In the context of managing common property natural resources, the authors drew attention to the great ethnic diversity that exists at small scales. They called into question the common rhetoric associated with 'community'-based approaches. Failure to recognise such diversity on the part of policies, governance systems and development planners can result in scale mismatches (Cumming *et al.*, 2006) between institutions and resource management, and in the disenfranchisement of some groups.

An overview of the historical and social dynamics on the edge was provided in Chapter 6, together with a broader construct of 'the edge' and the role of historical contingency in how the edge is viewed and used both positively and negatively. The edge, and its associated TFCAs, open up entrepreneurial opportunities, both legal and illegal, that may divert the TFCA project into unintended directions – a dynamic that may, in the end, prejudice both conservation and development for those living on the edge. The many competing claims for resources and space on the edge add to the potential for conflict and further marginalisation of people living within still ill-defined TFCAs.

Boundaries are characterised by asymmetries and Chapter 7 examined how these are shaped in TFCAs. Boundaries range from abrupt hard edges to gradients in natural resources that stimulate movements of wild and domestic animals, and people, to and fro along gradients in search of resources. The resulting asymmetries can result in sources and sinks and ecological traps. They can also result in social traps in the sense of breaking laws and social norms, with ensuing penalties. Examples of movements driven by resources gradients across the boundaries of TFPs include predators leaving parks to feed on livestock, elephants going in search of nutritious food in croplands, cattle seeking grass in under-stocked parks, people harvesting forestry products and bushmeat, and rustling cattle across national borders because of asymmetries in prices and markets. One of the future management challenges of TFCAs will be to mitigate the negative consequences of resource-driven movements of animals and people, either by manipulating the gradients across boundaries or removing excess animals where they aggregate in sinks.

Chapter 8 examined the consequences of animals and people crossing the protected area edge. These include conflicts between wildlife and humans that can result in death, injury, loss of crops and livestock, transmission of diseases and also the loss of endangered wild animals, and negative attitudes towards conservation and protected areas. Human–wildlife conflict can result in high direct costs to those living on the edge, as well as indirect costs associated with the time and energy invested to protect assets and avoid conflict. The spread of diseases across international boundaries is a major issue for some countries that seek to protect livestock and associated export markets. Current management of human–wildlife conflict within TFCAs, including disease transmission, is largely inefficient because of a lack of human and financial resources, inappropriate techniques or policies, and a frequent lack of consideration of local peoples' perceptions of risk and farmer-based solutions.

A key issue for people living on the edge is that of finding ways to meet the costs of conserving wild natural resources and living with large, and sometimes dangerous, animals. This was taken up in Chapter 9, which provided an overview of the major features of 13 TFCAs in Southern Africa and the wide range in their characteristics, including their size, climate, resource base, population densities, land uses, diseases and opportunities. A review of existing CBNRM enterprises indicated that, with the exception of Namibia, the returns to communities and households are generally too small to provide adequate incentives to retain wildlife as a form of land use. An exploration of opportunities that could be exploited within TFCAs indicated that considerable benefits could accrue to communities and households. However, inappropriate legal and policy frameworks effectively forestall such opportunities. Only in Namibia have nearly full resource access rights and benefits been devolved to communities. This has resulted in expanding wildlife populations, a steadily increasing number of registered community conservancies and increasing revenues to these communities' natural resource-based enterprises.

TFCAs and the promise of conservation *and* development

The edge thus emerges as a zone of competing claims and opportunities, following its earlier marginality. Its future may be one of increased conflict and marginalisation of its inhabitants. On the other hand, these areas may give rise to diverse natural resource-based enterprises that enrich the lives of those who live there. However, the bottom line is that unless those living on the edge benefit from conserving wild natural resources, TFCAs are unlikely to work as sustainable multiple use zones. But a failure to link conservation and development in TFCAs may not be without consequences. The long-term future of the core protected areas within TFCAs is likely to be compromised, if not threatened, unless those living on the edge support their continued existence. While conservation in TFCAs may function as a driver of development on the edge it may also depend upon it. What then, are the conservation and development objectives for TFCAs? What incentives are there for people to support TFCAs, and what are the constraints to realising the promise of development and conservation as presented at the beginning of this book?

TFCA objectives

The Memoranda of Understanding and the Treaties for TFCAs in Southern Africa are remarkably silent on the conservation *and* development objectives for these areas (Cumming, 2011). Some conservation objectives are, of course, implicit in attempts to define wildlife corridors, or the resettlement of people from newly defined protected areas. The maintenance or restoring of biodiversity and uninhabited protected areas apparently remain important conservation objectives for the core protected areas of TFCAs. To be sure, these objectives are not easily met. First, many of the core-protected areas are inadequately funded and face ongoing declines in several large mammals (e.g. Cumming, 2004; Chase, 2011). Second, ecologically effective and functional corridors are yet to be established between the core protected areas of TFCAs. Third, some protected areas remain settled, or have recently been partially settled by people with prior claims on the area (Milgroom, 2012; Mombeshora and le Bel, 2009). Fourth, achieving conservation objectives in (planned) interconnected protected areas across national borders requires joint planning and policy formulation at the overall scale of TFCAs and the synchronisation of policies across the countries involved.

Attempts to formulate objectives for the edge of the core protected areas, the locus of conservation *and* development, are very recent. Explicit conservation objectives for the edge have yet to be formulated, and the synchronisation of national policies for resource conservation in different TFCA countries remains an important first step to be made. Proposed development objectives focus on incentive structures for wildlife- and natural resource-based enterprises, and non-TFCA-specific development objectives such as enhancing food security through improved and diversified smallholder farming systems and market integration (Cumming, 2011).

Legal frameworks and incentives

Legal frameworks and peoples' rights to resources and to engage in influencing land use and resource access planning are generally weak in Southern Africa (Mohamed-Katetere, 2001). National interests may override international law, and even the Southern African Development Community (SADC) instruments, as exemplified by the recent suspension of the SADC tribunal following its ruling on land reform in Zimbabwe. Existing access rights to natural resources at national levels in most Southern African countries prejudice the development of effective natural resource-based enterprises. Policies and incentive structures are presently based on inadequate information about the full costs and implications of alternative land use options in TFCAs. The lack of sound spatial and temporal information on biodiversity, land use and human welfare (including the incidence and prevalence of diseases) in TFCAs throughout the region is also a major concern. This constrains conservation and development planning by resource managers and villagers (see Getz *et al.*, 1999). This is particularly important in relation to decisions on disease control strategies that are aimed at improving conditions for beef exports, at the expense of biological conservation, ecosystem services, and alternative land use options.

Much of the development focus has been on the integration of agriculture and natural resource management on the edge, the common ground of CBNRM. Yet TFCA policy and planning currently has neither a CBNRM strategy, nor a framework to deal with common problems in CBNRM, such as human–wildlife conflict and equitable distribution of benefits. Similarly, concerns have been raised about transboundary animal diseases affecting wildlife and livestock in TFCAs and beyond, but community-based animal health systems (e.g. Grahn and Leyland, 2005) have not been developed. Appropriate devolution of rights and responsibilities for natural resource management has been lacking, with the result that issues such as the lack of involvement of local people in resource management benefit capture by local elites have bedevilled CBNRM initiatives in TFCAs and elsewhere. Murphree (2000) captured the essence of the issue thus: 'In other words, robust jurisdictional devolution has not really been tried, reminiscent of G.K. Chesterton's comment on Christianity: "Christianity has not been tried and found wanting; it has been found difficult and not tried".'[1]

Constraints across institutional levels

Essentially TFCAs provide a new layer in the hierarchy of ecological scales and institutional levels that influence the lives of people living on the edge (Figure 10.1). Thus the constraints to conservation and development success operate across at least four levels. The first is that of international influences, laws and conventions. Inserted as a second level, TFCAs aim to link and create synergy across country boundaries. The third is the national level, at which international treaties and conventions are translated and implemented, and at which national management capacities become apparent. And finally the local level, within TFCAs at the wildlife–livestock–human interface, forms the fourth level with

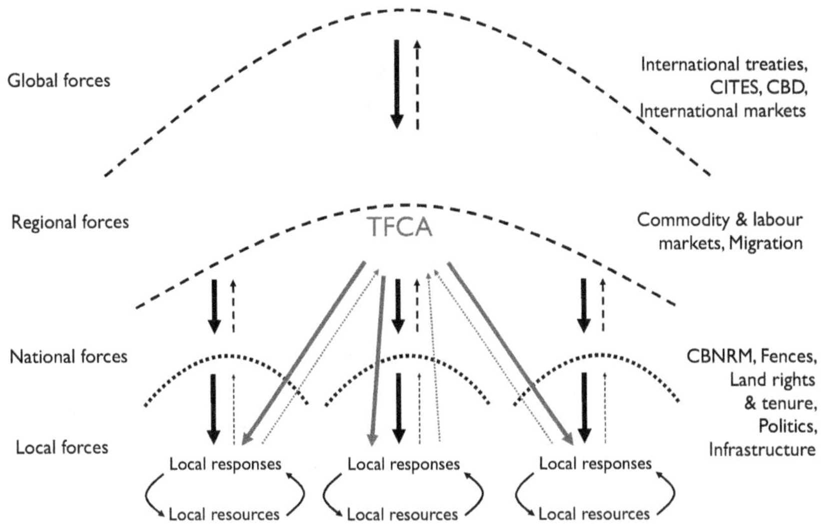

Figure 10.1 At the local level, responses are dependent on resources, but opportunities are constrained by pressures from above.

all the diversity and complexities described earlier. Factors at any one of these levels can prove to be binding constraints to successful outcomes at local levels. For example, KAZA carries close to 250,000 elephants. However, because international trade in ivory is presently banned by the CITES, communities within KAZA, and elsewhere, are unable to realise the benefits from ivory obtained from elephants that die, or are killed in problem animal control operations in protection of people and crops, in their areas. The full potential value of elephants to communities on the edge is therefore reduced. Control of animal diseases, and human–wildlife conflict, is generally regarded by local people as the responsibility of the state and its agents, because problems such as these have typically been managed through centralised command and control systems. Yet livestock keepers have their own perception of local priorities relating to diseases, which are seldom taken into account. Poor information flow from the higher echelons of government agencies excludes and alienates those on the ground from making positive contributions to development within TFCAs. Thus local ownership of problems and opportunities for innovation at this level are stifled.

International treaties, such as the Convention on Biological Diversity (CBD), the Convention on International Trade in Endangered Species (CITES), or the Terrestrial Animal Health Code of OIE (the World Organisation for Animal Health), operating at the global level, have a direct influence on responses at the national level (Figure 10.1). TFCAs influence international levels indirectly through influencing national policies and directly at local level where policies are implemented. Feedback from local and national levels is weaker than the constraints imposed from higher levels (Figure 10.1).

The advent of TFCAs thus demands devolution of international regulations to the level of TFCAs. Simultaneously, it demands alignment of approaches and policies of the countries that they encompass directly. But the effects of TFCAs are much wider than simply influencing countries with contiguous boundaries. For example, South Africa shares TFCAs with Namibia, Botswana, Mozambique and Zimbabwe, but also has to align approaches with Zambia, which shares TFCAs with all of these countries. This highlights the need for regional approaches to TFCAs. However, a pronounced diversity and heterogeneity within and between edge areas and TFCAs emerges strongly in the preceding chapters. This is apparent in biophysical, social, cultural and economic spheres and TFCA sizes ranging from 2,000 to 400,000 km^2, covering deserts to tropical forests, and having human population densities ranging less than 5 in parts of KAZA to more than 300 people per km^2 on the western boundary of KNP. Great disparities in wealth and capacity exist between the countries involved in TFCAs, with GNI per capita ranging from US$197 in Zimbabwe to US$13,204 in Botswana (*CIA World Factbook*, 2011). Regional and national cultural differences are mirrored at local scales, not only across national boundaries, but also within countries among districts, wards and villages. This local social and cultural heterogeneity may in turn be reflected in alternative and conflicting approaches to managing natural resources and emphasises the challenges facing the development of harmonised regional approaches to dealing with the 'edge'.

Local level constraints and opportunities

The remoteness and poor infrastructure in most of the TFCAs, which are located in border areas, has negative effects on local development and yet positive effects for conservation. Provision of services, such as medical centres and schools, and improvements in road access can encourage in-migration. But they could also reduce pressures on natural resources on the edge as they contribute to alternative livelihood opportunities, out-migration and reduced population growth. As development elsewhere in the world has shown, investments in health and education are key to stemming population growth. Poor infrastructure also constrains cross-border interchange between resource managers, while security regulations and cross-border travel restrictions also pose serious constraints.

Greater control over income from trophy hunting or harvesting of bushmeat would yield more benefits for local people, but could also attract more people to the area in question. Increased incomes from cotton production have driven in-migration, land clearance and loss of wildlife in the Mid-Zambezi Valley (Baudron *et al.*, 2011). These examples resonate with 'Jevon's paradox', which suggests that increases in the efficiency with which a resource is used leads to faster depletion of that resource (Jevons, 1866). It is clear that simply increasing incomes without building institutions to govern resource use is likely to work against conservation through development.

Local capacity for self-organisation and institution building for resource management is often weak due to the diverse interests and goals among local people as well as interference from higher levels. A lack of trust in local and national institutions means that there are high transaction costs and substantial time is needed for investment in local institutions to aid development.

Cross-scale issues

TFCAs are essentially complex social-ecological systems, undergoing constant change and adapting to social and biophysical drivers at international, national and local scales and levels. Cross-scale impacts, for example social factors such as global financial crises, affect development assistance and national and local prices of food. Climate change and increasing variability in rainfall affect crop production and livelihoods as well as wildlife habitats. National polices and laws govern access to natural resources, particularly wildlife, and the use and benefits to be derived from them. International conventions and obligations can also affect access to, and benefits to be derived from, wildlife, such as the example of a ban on international ivory trade mentioned above. Internationally imposed trade bans can represent a mismatch in scale between institutions and resources where national and local resource managers would, given appropriate legal frameworks and incentives, be better placed to sustainably manage and legally benefit from these resources. Scale mismatches between institutions and resource management are widespread and pervasive (Cumming *et al.*, 2006, 2012).

Clear examples of the successful devolution of resource management institutions to appropriate ecological and resource management scales are the granting of appropriate authority for wildlife to landholders in Namibia, South Africa and Zimbabwe (Bond and Cumming, 2006; Suich *et al.*, 2009). The devolution of responsibility for the management of wildlife to community conservancies in Namibia (Weaver *et al.*, 2011) is a clear example involving common property resources. Within TFCAs, such devolution will need to be matched with integration at the TFCA and regional level.

TFCAs, globalisation and increased vulnerability

Opportunities for local development cannot be seen in isolation from wider developments in Southern Africa and across the world. TFCAs are part of an increasingly interconnected, globalised world. Achieving conservation objectives within TFCAs may thus be highly dependent on the creation of opportunities elsewhere to relieve land and resource pressures on the conservation edge. Pressure on land will surely increase unless the increase in the human population is stemmed and alternatives to living in remote rural areas are created. Improvements in wealth, health and education are key to reducing population growth rates throughout the world indicating that conservation and development have to proceed hand-in-hand.

TFCAs may result in greater ecological and socio-economic connectivity across large landscapes in order to achieve desired conservation, development and political goals. Increased connectivity can, however, come at a cost, such as increased exposure to a variety of unwanted, or unanticipated, processes. These may include greater vulnerability to transboundary infectious diseases, increased smuggling of contraband, illegal migrants and exposure to social and economic instability in neighbouring states. Modularity is an important property of resilient systems (Levin, 1999; Brock and Carpenter, 2007) but where the balance between establishing connectivity and maintaining modularity in TFCAs resides is a problem yet to be explored. An important aspect of this problem is that of maintaining ecosystem services at scales from the local to the global and who pays for them.

One option is to allow greater sharing of the costs and benefits of conserving biodiversity across the globe through Payment for Environmental Services (PES). Carbon markets for polluting nations to mitigate climate change (see Chapter 9) are one such option. Similar schemes are under discussion that would recognise the right of local people to live in biodiversity rich areas, but aim to pay them to manage such landscapes in a benign or low-intensity manner (Wunder *et al.*, 2008). CAMPFIRE could be considered as an early example of a PES scheme that relied on incomes from trophy hunting. Other approaches involve compensation payments for predation by carnivores (Dickman *et al.*, 2011). Development of new PES schemes requires substantial initial investments to establish local institutions to manage them and to make them operational, but such schemes may allow some imbalances of local costs and global benefits to be redressed.

Common property regimes and self-organisation

The core issues of conservation and development in TFCAs revolve around the management of common property regimes, resource access rights and self-organisation at appropriate scales. Population to resource ratios and expanding human populations are also vitally important. In part, Mudemi (see Chapter 9) summed up the common property dilemma with his observation that 'the problem is that the wildlife is ours but the cattle are mine'. What is missing from his observation, however, is the realisation that the grass, on which the cattle feed, is 'ours', i.e. grazing is common property in communal land tenure systems and the basis of Hardin's 'tragedy of the commons' metaphor.

State-protected areas in the region are, for the most part, surrounded by communal lands in which natural resources are managed under common property regimes. The predominant focus of common property scholarship and CBNRM has been single resources such as water, forests, grazing or wildlife (e.g. Ostrom *et al.*, 2007). But TFCAs typically deal with multiple, diverse and interconnected common pool resources, including fugitive resources such as wildlife, which adds to the complexity of managing common pool resources (Ostrom, 2009).

The prevailing advocacy of single recipes, or mantras, such as devolution, CBNRM, Integrated Conservation and Development Projects (ICDPs), privatisation, public–private–community partnerships and so on, fall into the category of mostly failed panaceas to which Ostrom *et al.* (2007) so ably drew attention. Ostrom and her colleagues extended their critique by outlining a framework that allows a deeper understanding of common property regimes at the local level

(Meinzen-Dick, 2007; Ostrom, 2009). How do the issues confronting TFCAs fit within Ostrom's framework and the ten necessary components[2] for common property regimes to work? And to what extent have TFCAs paid attention to Berkes' (2007) essay on CBNRM in a globalised world? Or to Armitage *et al.*'s (2009) and Waltner-Toews *et al.*'s (2008) approaches to ecosystem management and Giller *et al.*'s (2008) suggestions of how to tackle the thorny issues of competing claims on land and natural resources. The answer, at least at an official level, is little if at all. If anything, TFCAs present us with several 'wicked problems' (Rittel and Webber, 1973) that are unlikely to be resolved readily.

Conclusion

We show that opportunities for development for people living on the edge are influenced by a multitude of factors from the global to the local level. There is clearly no single silver bullet solution. Peace parks, and the TFCAs spun off from them, were seen as an opportunity to achieve development through conservation. It is clear that development and conservation are difficult to achieve by focusing on local development at the edge. This suggests the need to move to a paradigm of development *for* conservation to provide attractive alternatives to encourage migration away from the edge. Such development is necessarily multi-scale and multi-level in nature, but requires local gravitation in order for people on the edge to meaningfully participate in their own future on the protected area edge. At present the benefits of living on the edge are outweighed by the costs of human–wildlife conflict and spread of livestock diseases. By addressing scale issues, key policy and legal constraints, capacity to self-organise (empowerment!) at local level and at appropriate scales, it may be possible to achieve effective conservation and development on the edge and allow Southern African TFCAs to play a central role in this. Wild nature remains emblematic for Africa, and the global support for conservation is strong. Yet the burden of living with wildlife rests firmly on the shoulders of local people. Is it not time for the global community to bear the true costs of conservation and offer realistic alternatives for people living on the edge?

Notes

1 Murphree's footnote to his quotation of Chesterton's comment was, 'This quote from Chesterton is embedded in my memory from a reading now 40 years old. I cannot cite the precise reference, but if pushed could do so. Better still, let the reader peruse Chesterton's writings, which would be a salutary experience on its own'. The original of Chesterton's text was 'The Christian ideal has not been tried and found wanting; it has been found difficult and left untried.'

2 See Chapter 9 where these are listed and discussed.

References

Armitage, D.R., Plummer, R., Berkes, F., Arthur, R.I., Charles, A.T., Davidson-Hunt, I.J., Diduck, A.P., Doubleday, N.C., Johnson, D.S., Marschke, M., McConney, P., Pinkerton,

E.W. and Wollenberg, E.K. (2009) 'Adaptive co-management for social-ecological complexity', *Frontiers in Ecology and Environment'*, vol. 7, no. 2, pp. 95–102

Baudron, F., Corbeels, M., Anderson, J.A., Sibanda, M. and Giller, K.E. (2011) 'Delineating the drivers of waning wildlife habitat: the predominance of cotton farming on the fringe of protected areas in the Mid Zambezi Valley; Zimbabwe', *Biological Conservation*, vol. 144, no. 5, pp. 1481–1493

Berkes, F. (2007) 'Community-based conservation in a globalised world', *Proceedings National Academy of Sciences*, vol. 104, pp. 15188–15193

Bond, I. and Cumming, D.H.M. (2006) 'Wildlife research and development', in M. Rukuni, P. Tawonezi, C. Eicher, M. Munyuki-Hungwe and P. Matondi, (eds) *Zimbabwe's Agricultural Revolution Revisited*, 2nd ed., University of Zimbabwe Publications, Harare, pp. 477–496

Brock, W.A. and Carpenter, S.R. (2007) 'Panaceas and diversification of environmental policy', *Proceedings National Academy of Sciences*, vol. 104, pp. 15204–15211

Chase, M. (2011) 'Dry season fixed-wing aerial survey of elephants and wildlife in northern Botswana', Report from Elephants Without Borders, Kasane, Botswana and Department of Wildlife and National Parks, Gaborone

CIA World Factbook (2011). Available at: www.cia.gov/library/publications/the-world-factbook/ (accessed July 2011)

Cumming, D.H.M. (2004) 'Performance of parks in a century of change', in B. Child (ed.), *Parks in Transition: Biodiversity, Development and the Bottom Line*, Earthscan, London, pp. 105–124

Cumming, D.H.M. (2011) 'Constraints to conservation and development success at the wildlife–livestock–human interface in Southern African transfrontier conservation areas: a preliminary review', Technical Report to the Wildlife Conservation Society (WCS). Available at: www.wcs-ahead.org/kaza/kaza_constraints_to_conservation.html

Cumming, G.S., Cumming, D.H.M. and Redman, C.L. (2006) 'Scale mismatches in social-ecological systems: causes, consequences, and solutions', *Ecology and Society*, vol. 11, no. 1, p. 14. Available at: www.ecologyandsociety.org/vol11/iss1/art14/

Cumming, G.S., Olson, P., Chapin III, F.S. and Holling, C.S. (2012) 'Resilience, experimentation, and scale mismatches in social-ecological systems', *Landscape Ecology*. Available at: DOI 10.1007/s10980-012-9725-4

Dickman, A.J., Macdonald, E.A. and Macdonald, D.W. (2011) 'A review of financial instruments to pay for predator conservation and encourage human–carnivore coexistence', *Proceedings of the National Academy of Sciences, USA*, vol. 108, pp. 13937–13944

Getz, W.M., Fortmann, L., Cumming, D.H.M, Toit, J. du, Hilty, J., Martin, R., Murphree, M., Owen-Smith, N., Starfield, A.M. and Westphal, M.I. (1999) 'Sustaining natural and human capital: villagers and scientists', *Science*, vol. 283, pp. 1855–1856

Grahn, R. and Leyland, T. (2005) 'Complementarity between community-based animal health delivery systems and community-based wildlife management? An analysis of experiences linking animal health to conflict management in pastoralist areas of the horn of Africa', in S.A. Osofsky, S. Cleaveland, W.B. Karesh, M.D. Kock, P.J. Nyhus, L. Starr and A. Yang (eds), *Conservation and Development Interventions at the Wildlife/Livestock Interface: Implications for Wildlife, Livestock and Human Health*, IUCN, Gland, pp. 133–139

Giller, K.E., Leeuwis, C., Andersson, J.A., Andriesse, W., Brouwer, A., Frost, P., Hebinck, P., Heitkönig, I., Ittersum, M.K. van, Koning, N., Ruben, R., Slingerland, M., Udo, H., Veldkamp, T., Vijver, C. van de, Wijk, M.T. van and Windmeijer, P. (2008) 'Competing claims on natural resources: what role for science?', *Ecology and Society*, vol. 13, no. 2, p. 34. Available at: www.ecologyandsociety.org/vol13/iss2/art34/

Jevons, W. J. (1866) *The Coal Question: An Inquiry Concerning the Progress of the Nation, and the Probable Exhaustion of Our Coal-Mines*, 2nd ed., Macmillan and Co., London

Levin, S.A. (1999) *Fragile Dominion: Complexity and the Commons*, Perseus Publishing, Cambridge, MA

Milgroom, J. (2012) 'The elephants of democracy: an unfolding process of resettlement in the Limpopo National Park', Ph.D. thesis, Wageningen University, Wageningen

Meinzen-Dick, R. (2007) 'Beyond panaceas in water institutions', *Proceedings National Academy of Science*, vol. 104, pp. 15200–15205

Mohamed-Katerere, J. (2001) 'Review of the legal and policy framework for transboundary natural resource management in Southern Africa', Paper No. 3, *IUCN-ROSA Series on Transboundary Natural Resource Management*, IUCN-ROSA, Harare

Mombeshora, S. and Le Bel, S. (2009) 'Parks–people conflicts: the case of Gonarezhou National Park and the Chitsa community in south-east Zimbabwe', *Biodiversity and Conservation*, vol. 18, no. 10, pp. 2601–2623

Murphree, M.W. (2000) 'Multiple boundaries, borders and scale', 8th Biennial Conference of the International Association for the Study of Common Property, Bloomington, IN

Ostrom, E. (2007) 'A diagnostic approach for going beyond panaceas', *Proceedings of the National Academy of Sciences*, vol. 104, pp. 15181–15187

Ostrom, E. (2009) 'A general framework for analyzing sustainability of social-ecological systems', *Science*, vol. 352, pp. 419–422

Ostrom, E., Janssen, M.A. and Anderies, J.M. (2007) 'Going beyond panaceas', *Proceedings National Academy of Sciences*, vol. 104, pp. 15176–15178

Rittel, H. and Webber, M. (1973) 'Dilemmas in a general theory of planning', *Policy Sciences*, vol. 4, pp. 155–169

Suich, H., Child, B. and Spenceley, A. (eds) (2009) *Evolution and Innovation in Wildlife Conservation: Parks and Game Ranches to Transfrontier Conservation Areas*, Earthscan, London

Waltner-Toews, D., Kay, J.J. and Lister, N-M.E. (2008) *The Ecosystem Approach: Uncertainty and Managing for Sustainability*, Columbia University Press, New York

Weaver, L.C., Hamunyela, E., Diggle, R., Matongo, G. and Pietersen, T. (2011) 'The catalytic role and contributions of sustainable wildlife use to the Namibia CBNRM program', in M. Abensperg-Traun, D. Roe and C. O'Criodain (eds), *CITES and CBNRM. Proceedings of an international symposium on 'The relevance of CBNRM to the conservation and sustainable use of CITES-listed species in exporting countries'*, Vienna, Austria, 18–20 May 2011, IUCN, Gland and IIED, London, pp. 59–70

Wunder, S., Engel, S. and Pagiola, S. (2008) 'Taking stock: a comparative analysis of payments for ecosystem services programs in developed and developing countries', *Ecological Economics*, vol. 65, pp. 834–852

Index

Page numbers in *italic* indicate relevant figures and tables; **bold** page numbers indicate boxed text; capitalizations, word forms, hyphenations and spellings follow the majority usage.